Star Vistas

Star Vistas

A Collection of Fine Art Astrophotography

Greg Parker and Noel Carboni

Springer

Printed on acid-free paper.

Library of Congress Control Number: 2008942037

Greg Parker MobiusLtd@aol.com
http://www.NewForestObservatory.com

Noel Carboni NCarboni@att.net
http://www.ProDigitalSoftware.com

ISBN: 978-0-387-88435-6

Printed in the United States of America.

9 8 7 6 5 4 3 2 1

springer.com

We dedicate Star Vistas to the memory of Sir Arthur C. Clarke, 1917-2008, techno-visionary and a great writer of science fiction.

We shall miss you and we'll watch for you "out there."

- Greg Parker, Noel Carboni

Forewords

I must admit right away that my favourite celestial objects have always been the odd assortment known as the Solar System - our local neighbourhood in the vast Universe. In my childhood and youth, I spent countless hours observing the Moon, Mars and Jupiter using a home-built telescope. For the best magnification I could achieve, the stars were no more than bright sparks in the night sky.

How things have changed in just a few decades! Space exploration has expanded our knowledge and understanding of the heavens as never before. We have learnt more about the universe in the first 50 years of the Space Age than we did for several centuries since Galileo turned a refracting telescope on the night sky circa 1609. Telescopes in Earth orbit now peer deeper in space and farther back in time than he could possibly have imagined.

Confronted by these advances, amateur astronomers - the driving force behind the study of heavens for much of history - have had to reinvent themselves. Today's 'amateur' is, in fact, a serious professional armed with the latest tools of optics and computers to capture, analyse, catalogue and disseminate information and images of the night sky (and day sky!) They complement, rather than compete with, the work of major observatories and research scientists. Some areas of study, such as the search for near Earth objects - asteroids and comets - still rely heavily on the efforts of dedicated amateurs.

Dr Greg Parker is one such serious amateur. He is a professor of photonics and runs his own company by day. But his hobby and passion is to operate his own New Forest Observatory by night, from where he has been taking breath-taking images of deep sky objects for the past few years. He shares the best of these images through his website, and now, in this book.

In spite of my preoccupation with the Sun's suburbs, I am very interested in understanding just what's going in the far corners of the universe. I have suggested that supernovae are industrial accidents of advanced civilisations. Or, as I wondered in one of my best known short stories 'The Star', they may be inter-stellar beacons or signals of superior beings...

I have always found deep sky objects sparking my imagination, sometimes in the wildest directions. I hope the images in this book will have a similar effect on you.

- Sir Arthur C Clarke

Colombo, Sri Lanka
September 19, 2007

There was a time, not so very many decades ago, when good astronomical photography could be carried out only by professional observers using large telescopes in high-altitude decidedly inaccessible locations. Things are very different today. Greg Parker does hold a good degree in astronomy, but this is certainly not his full-time occupation; he is in fact a professor of photonics. He does not use elaborate, ruinously expensive equipment; his telescope is simply a conventional 11-inch reflector. And he is not inaccessible; the New Forest is a great deal easier to reach than, say, the northern part of the Atacama Desert, or the top of Mauna Kea in Hawaii. He also has to contend with the erratic British climate, where long sequences of clear nights are depressingly rare.

Yet the deep-sky images presented in this book are far better than anything which could have been produced when I began my "Sky at Night" television series in 1957. Recently I took one of his images – it happened to be of the Rosette nebula – and compared it with a picture of the same object taken in 1960 with the 200-inch Hale reflector at Palomar, then much the largest and most powerful telescope in the world. The Palomar view was less sharp, less striking and much less detailed.

There are two reasons for these changes. One is that we have entered the Electronic Age, and photography has started to seem very old-fashioned. (Be honest. Do you still use a film camera, or have you, like me, "gone digital?") The other is the skill of the user of a "back-garden" telescope. Greg Parker is one of a select group demonstrating this in no uncertain manner. He takes the images; they are processed by his American collaborator, Noel Carboni – and here is a selection of the results. The pictures are of real scientific value, and they are also works of art. You will enjoy them, and you will learn a great deal from them – as I did.

This is a book deserving an honoured place in every scientific library, amateur or professional. I have no doubt that it will have a wide circulation, and I am delighted to join my friends Arthur Clarke and Brian May in recommending it to you, together with my sincere congratulations to both authors.

As friends and astro colleagues of Prof. Greg Parker and Noel Carboni, we are all delighted that this book of their fine work has come to completion. Those of us over 30 (!) grew up enthralled by the analogue images of the night skies captured on film by the biggest telescopes of the day. Now, in the digital age, a few dedicated experts have been able to use the new technology to secure images of astronomical subjects which surpass anything which have been seen before.

Greg Parker uses a relatively modest telescope to capture light from these remote objects. But by cooling his CCD detector to reduce noise, and by perfect tracking and skilful manipulation, he is able to capture just the right photons to provide the building blocks for these wonderful new pictures. For any given field of view, instead of making a single exposure, Greg accumulates many images of exactly the same area of sky, often over many hours, using the optimum exposure to maximise signal-to-noise. These images then travel through cyber-space into the hands of Noel Carboni, who employs his own set of computer processes to assemble this set of data files into one image, of stunning quality.

These photographs are exceptional in their 'transparency' – they are free of the 'spread' which film photography imposed on bright objects, obscuring the faint objects around them. Hence, in these views, we are seeing hitherto unseen depths in many of the star-fields. I was fortunate to make Greg's acquaintance a few years ago, and many Parker-Carboni images have appeared on Brianmay.com, as well as in our book, Bang! The Complete History of the Universe, and on the Bang! website (Banguniverse.com).

Parker/Carboni have effectively enabled Nature to paint with a new brush, and to me the special thrill of some of these pictures is that we find ourselves looking at a naked-eye star-field which is quite familiar to us (my favourite being the 'Belt' of Orion) – but now we see spectacular coloured multitudes of stars, and billowing nebulosity which we never dreamed was there. These glories have always been right in front of us, yet not visible until now – it is as if we are seeing the sky for the first time with god-like eyes.

The wonders of the Universe are here!! Enjoy!

Preface

An international Internet collaboration between the authors of *Star Vistas* began in 2005. The computer and Internet-enabled fusion of my imaging skills and Noel's image-processing prowess has culminated in the striking astro images you will find in this book.

I started deep-sky imaging in November 2004, having just acquired a Starlight Xpress SXV-H9C one-shot color CCD camera. This camera provided 1.4 megapixels and coupled to a Starizona Hyperstar lens assembly it turned an optically slow 11-inch f/10 Celestron SCT telescope into a powerful f/1.85 astrograph. The high optical speed of the Hyperstar system meant that low-noise astronomical images could be acquired very quickly, and the shorter exposure times had the additional advantage that telescope mount drive precision was not as critical as for longer exposures. It was clear to me how to progress over time to improve my data acquisition, and in these early days when I also processed the images digitally, it was also becoming clear how to capture good data to minimize the image processing required. However, I was never particularly excited by the image processing aspect of deep-sky astrophotography, which can take a great deal of time, so it came as a great relief when a friend of mine, Bud Guinn, introduced me to the "Our Dark Skies" [ODS] Internet forum.

Noel was known on the ODS forum for transforming peoples' astronomical images into stunning deep-sky works of art using his Astronomy Tools software for Photoshop and his image processing skills, and he does some of his own astrophotography with digital SLR cameras. Noel worked his magic on one of my early images; it was a two-frame mosaic of M33, the Pinwheel galaxy in Triangulum, and he created the most striking image to come out of my observatory up to that time. The partnership was made! Noel realized my color camera and Hyperstar setup produced surprisingly good raw data to work with, and at last I had found an image-processing expert to help with the part of the image creation procedure I liked least.

The result of this Internet collaboration is the book you now hold in your hands. So far the authors have only communicated via the Internet; we have not yet met nor telephoned one another. It is truly remarkable how computers and the Internet have not only enabled communications and business but also allowed such long-distance collaborations – and friendships – to thrive.

Consider this: The *Star Vistas* image of the Quasar APM08279+5255, which lies at a distance of almost 13 billion light-years, was taken in the back garden of a home in the New Forest in Hampshire, U.K. The acquired data was sent via the Internet to Noel Carboni in the United States, who created the image you will find in this book. Photons from near the beginning of time itself – captured in a suburban observatory, transformed digitally halfway around the world, and presented to you in the pages of *Star Vistas!*

- Greg Parker, 2008

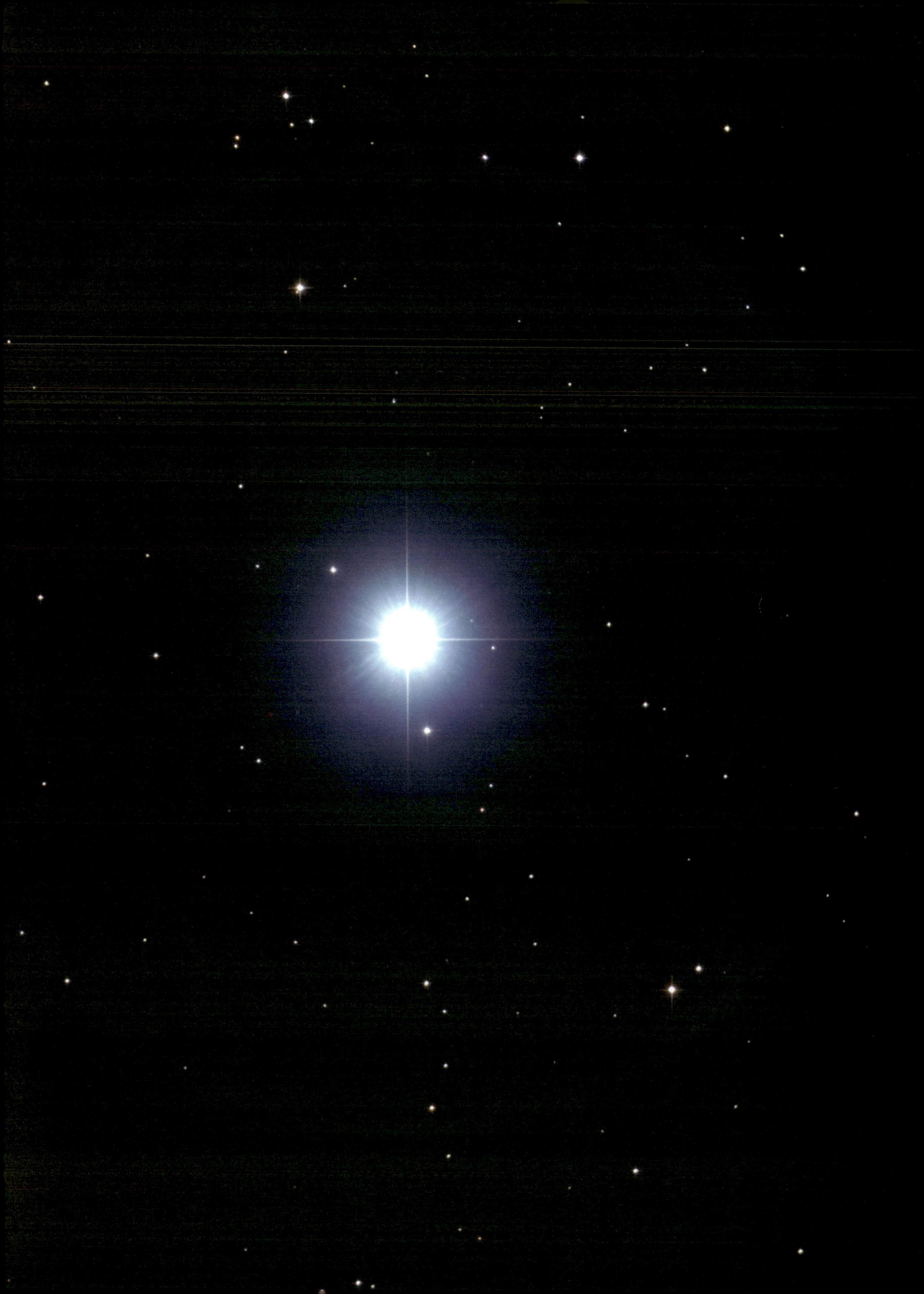

< Previous Page

Vega

The brilliant blue-white A0 V main sequence star Vega shines like a beacon against a stellar background in the Lyra constellation.

Vega is the 5th brightest star in the entire sky and can be seen from almost anywhere in the summer sky as part of the Summer Triangle of very bright stars, including Deneb and Altair at the other vertices. Vega is only about 25 light-years from Earth and is a relative youngster at only 500 million years of age. At twice the mass of our Sun, it is 50 times as luminous.

Though stars may appear in astro images as white circles or balls, they are really just very bright points of light, as even the most powerful telescopes can only barely resolve the very closest and largest stars into disks.

Vega image created from 3 minutes total exposure time using a Canon EOS-20D digital SLR at the Cassegrain focus of a 10-inch Meade LX200 GPS UHTC 2500mm f/10 Schmidt-Cassegrain telescope.

Opposite >

SH2-155 - The Cave Nebula

Ionized hydrogen glows red in the energetic light from stars in this part o large, complex region of nebulae strewn along the plane of our Milky Wa the constellation Cepheus.

Known as Sharpless 2-155 and Caldwell 9, the Cave Nebula gets its na from the bright curved region at the center of this image, which would l more like a cave if the exposure weren't quite so deep and the dark a appeared utterly black.

Cave Nebula image created from 2 hours total exposure time using Starlight Xpress SXV-H9C color (red, green, blue) camera on an 11" f/1 Starizona Hyperstar-equipped Celestron C11 Schmidt-Cassegrain telesc at prime focus (i.e., without secondary mirror in place).

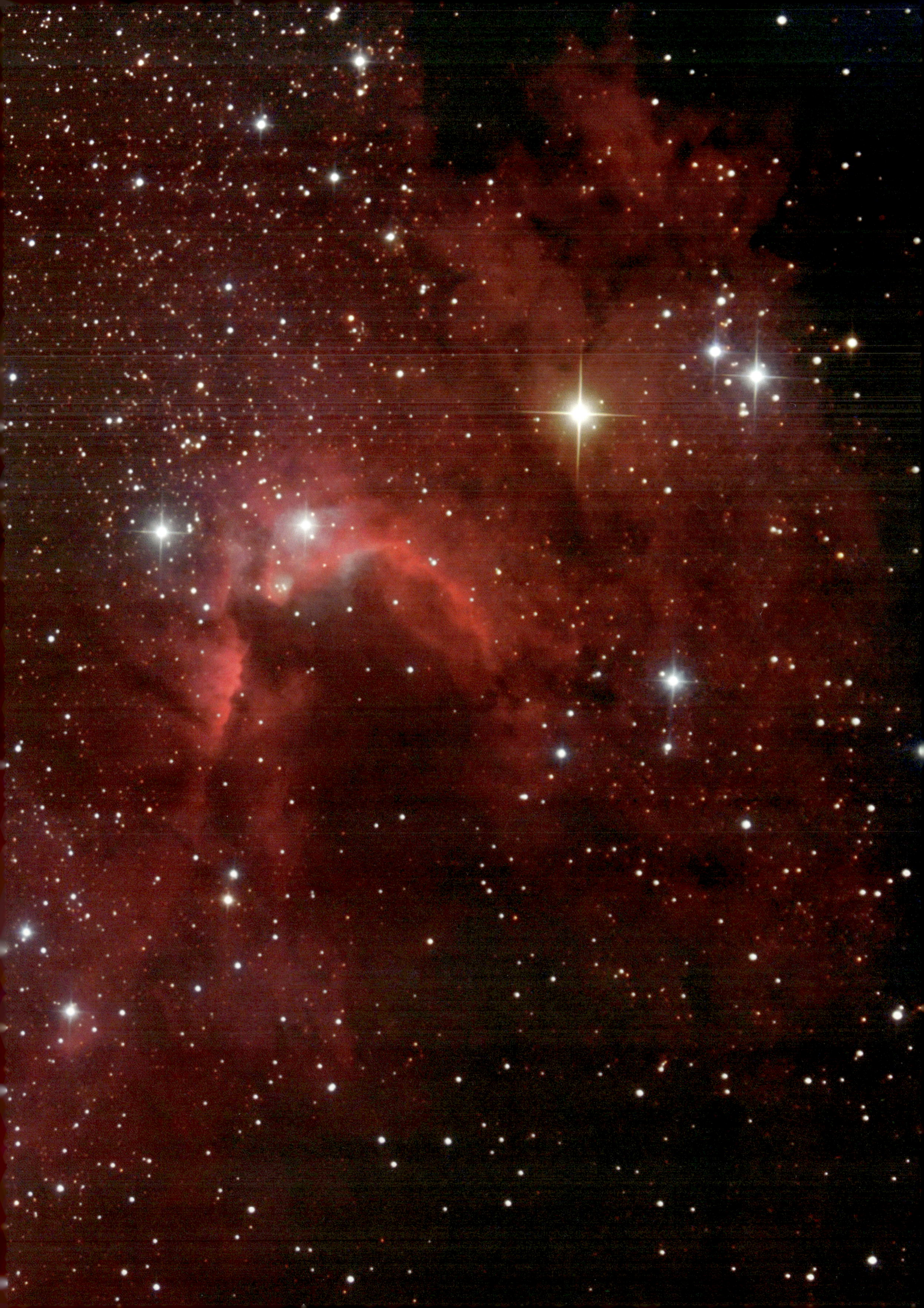

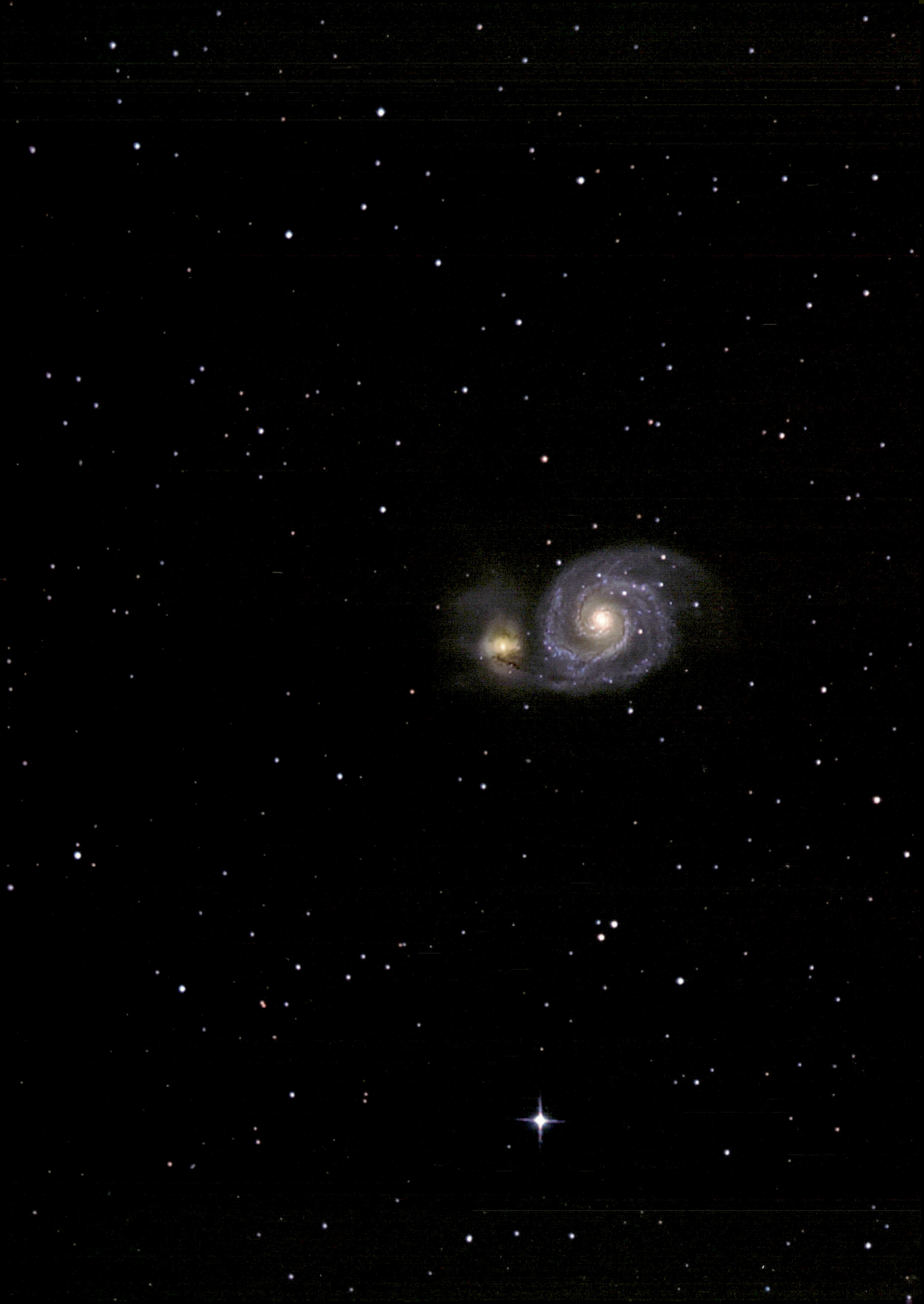

M51 - Interacting Galaxies

Galaxies are huge "island universes" millions of light-years away and containing billions of stars, often visible behind fields of stars that are much closer within our own Milky Way galaxy.

Discovered by Charles Messier in 1773, Messier's entry M51 (item number 51 in Messier's famous catalog of nebulae and star clusters) appears to describe the pair of galaxies, also known separately in the New General Catalog as NGC5195 and NGC5194. Though the larger appears to be in front of the smaller, many scientists believe they are near one another and interacting gravitationally.

A faint celestial thumbs up (or hitchhiker's thumb) seems to emanate from NGC5195 on the left. In fact these hazy, dim protrusions are clouds of millions of stars that have been thrown out of the galaxies due to their gravitational interaction with each other. The hot young blue stars in the arms of the spiral, and even the strong spiral structure itself, are considered indications of galactic interaction, as interstellar gas has been disturbed and compressed by gravity into forming hot new stars.

Seen easily with a small telescope under dark skies in the constellation Canes Venatici, not everyone agrees on the distance to these galaxies, but most put the figure between 20 and 40 million light-years from Earth.

M51 image: 1 hour 42 minutes RGB exposure, SXV-H9C / Hyperstar.

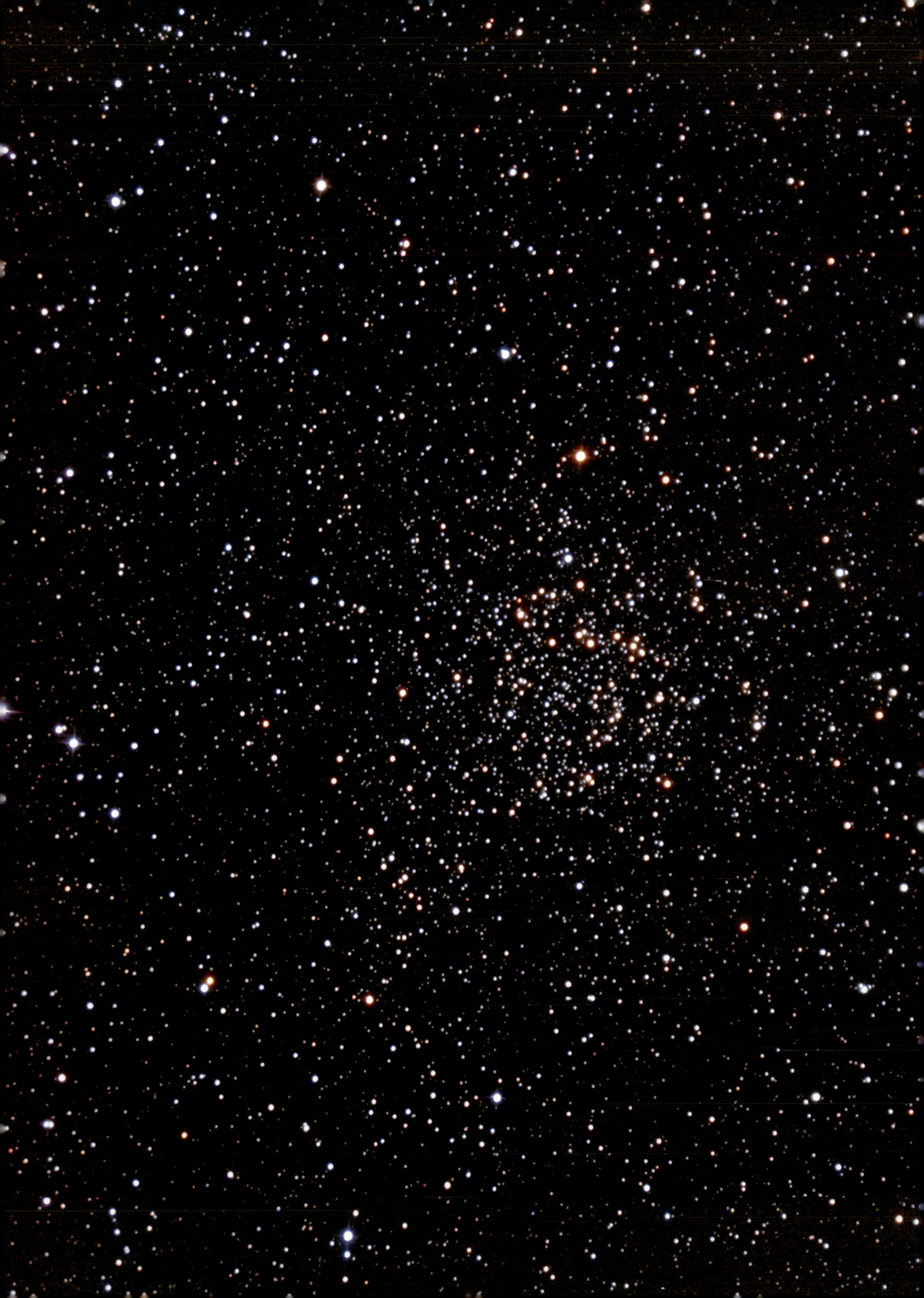

pposite

GC7789 – Open Cluster

pen cluster consists of stars all formed from the same cloud of gas at ut the same time. Measurements show them all moving in nearly the e direction at the same speed.

,000 light-years distant in our own Milky Way Galaxy in the constellation iopeia, this star cluster is beginning to show its age. It's about 1.6 billion s old, and the larger main sequence stars have all but exhausted the ogen fuel in their cores, blooming out into red giants. That's the same that will befall our Sun in another 5 billion years.

7789 image: 33 minutes RGB exposure, SXV-H9C / Hyperstar.

Next Two Pages >

M45 – The Pleiades

A reflection nebula is composed of interstellar dust of such a size that it scatters shorter wavelengths of light (i.e., blue) very effectively.

The Pleiades are stars in an open cluster that is inside a reflection nebula. In the northern hemisphere winter skies they form a well-known naked-eye constellation that looks a bit like a question mark or perhaps a tiny dipper. Known by many different names, the Pleiades were called Freya's Hens by the Vikings and Subaru by the Japanese, just to list a few.

M45 image stitched from several frames totaling 1 hour 14 minutes RGB exposure time, SXV-H9C / Hyperstar.

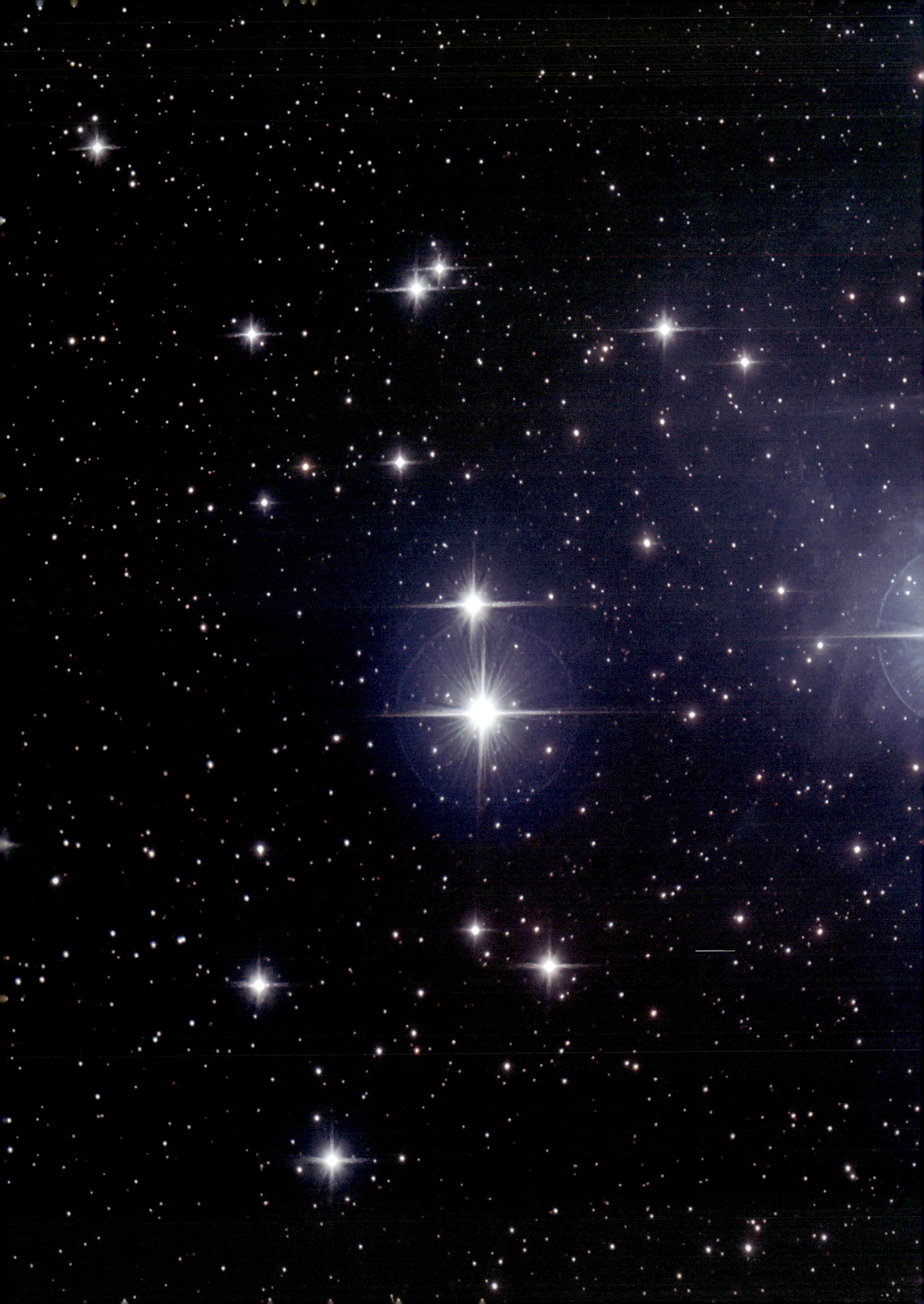

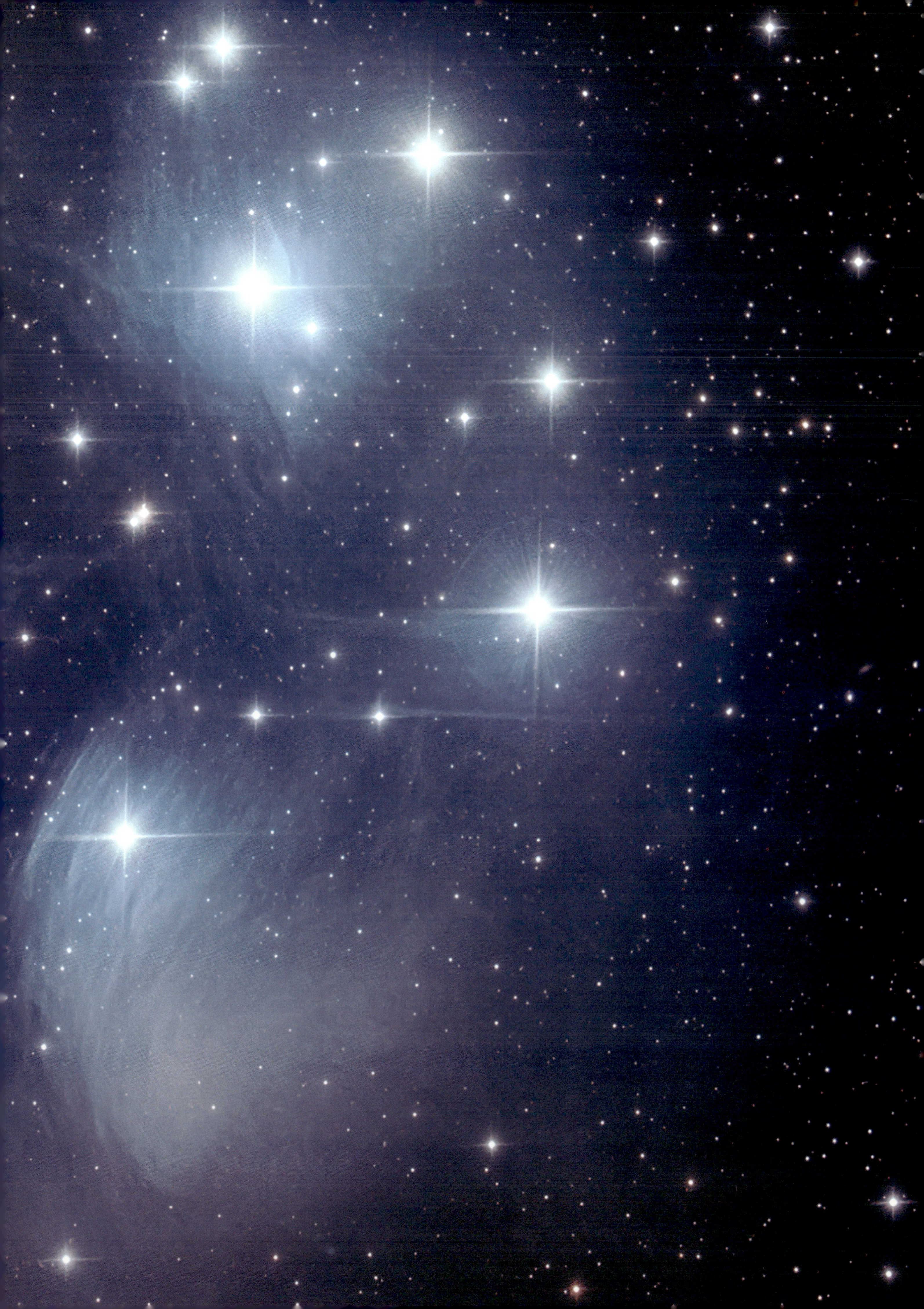

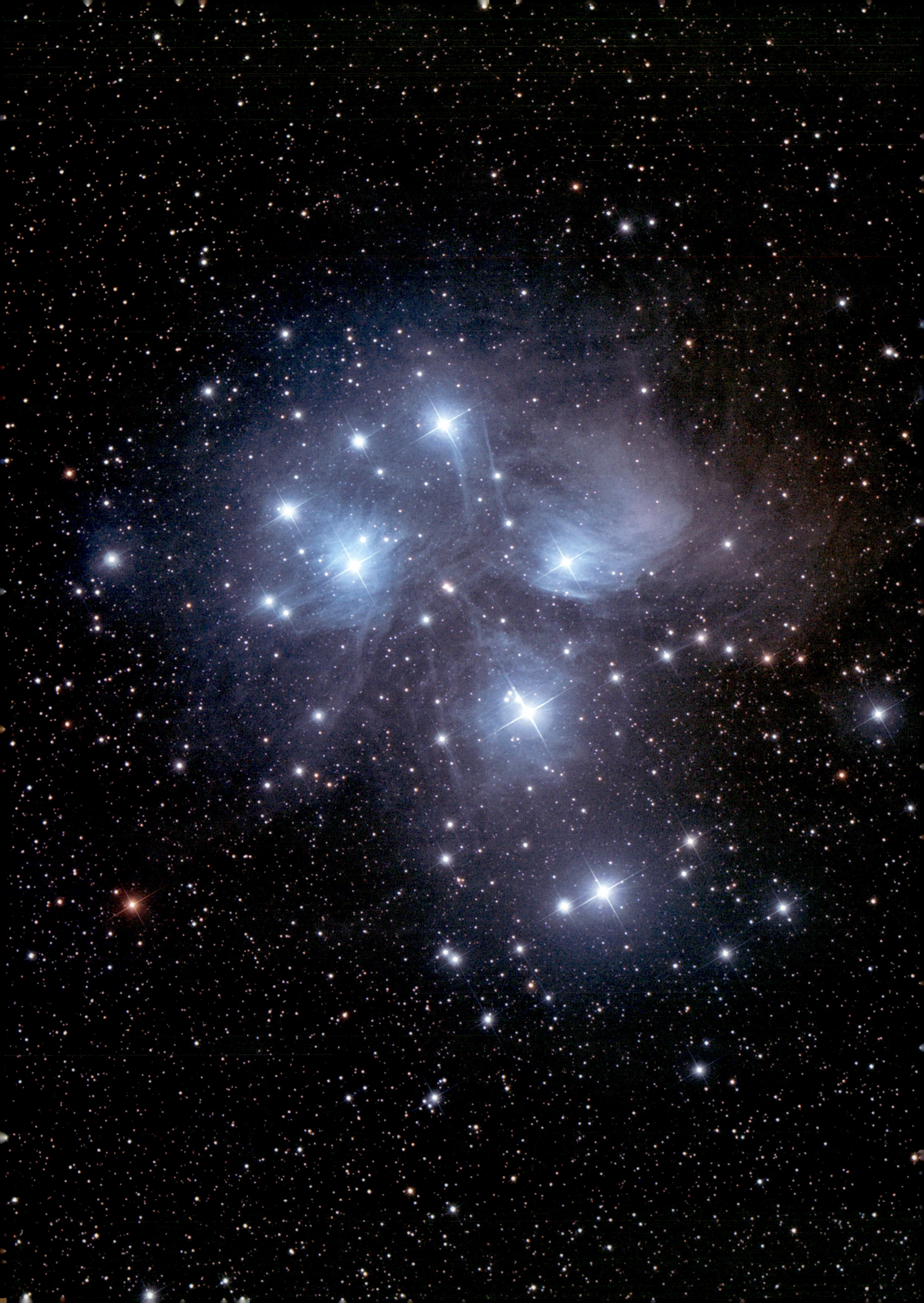

A Deeper, Wider View of the Pleiades

Showing the same cluster and nebula as on the previous two pages, this wider and slightly deeper view of the Pleiades in the constellation Taurus reveals that the reflection nebula extends out quite a distance, with a hint of a reddish hue toward the upper-right.

The dust being illuminated by the stars of this cluster was once thought to be a remnant of the material from which the cluster stars were originally formed, but scientists now believe that the stars are just passing through an unrelated interstellar cloud.

The Pleiades cluster lies at a distance of a little over 400 light-years. In astronomical terms it is relatively close to the Earth.

Pleiades wide-field image created from 5 hours 48 minutes total exposure time using a Starlight Xpress SXVF-M25C RGB camera on a Takahashi Sky-90 telescope with f/4.5 focal length reducer/field corrector.

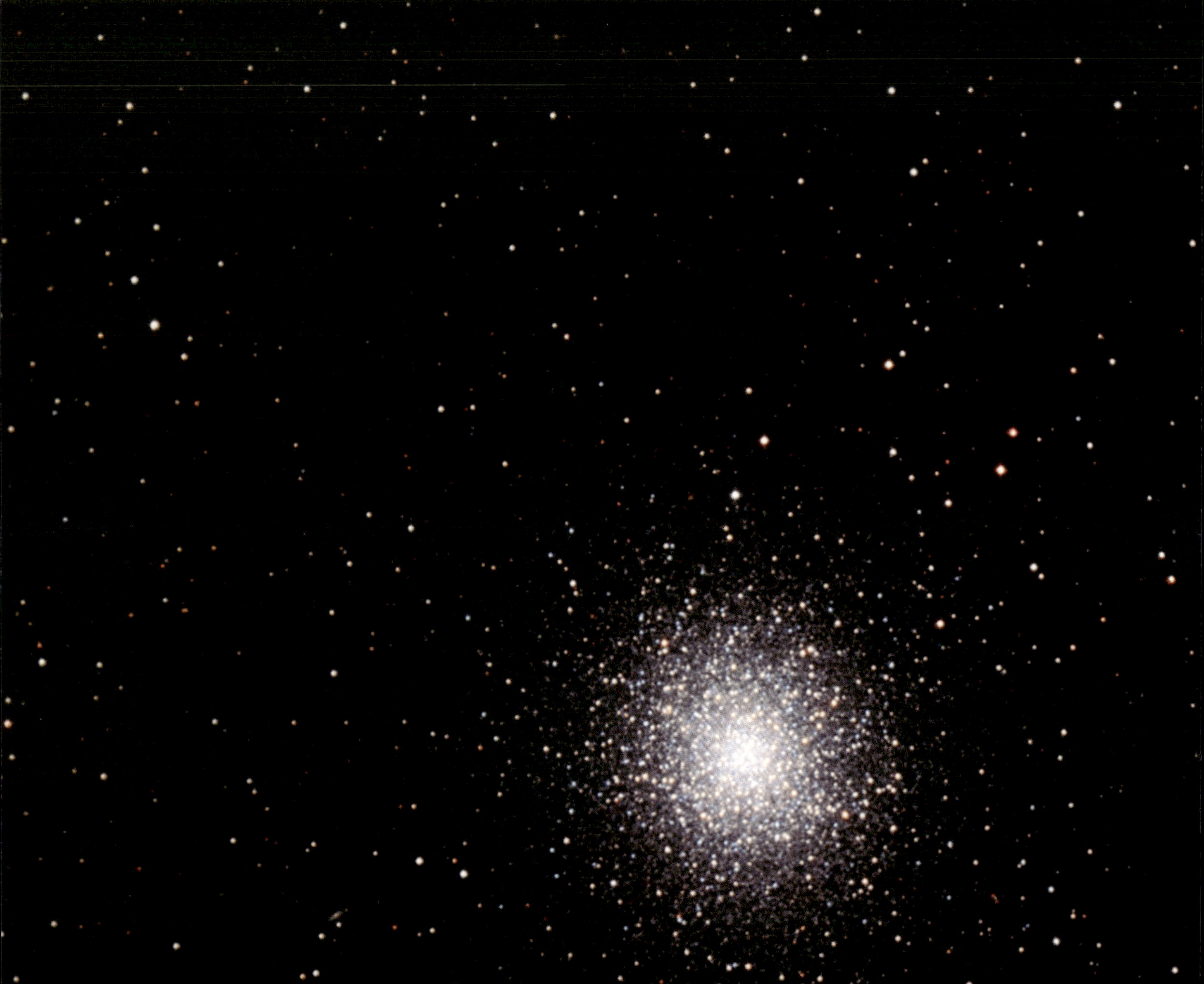

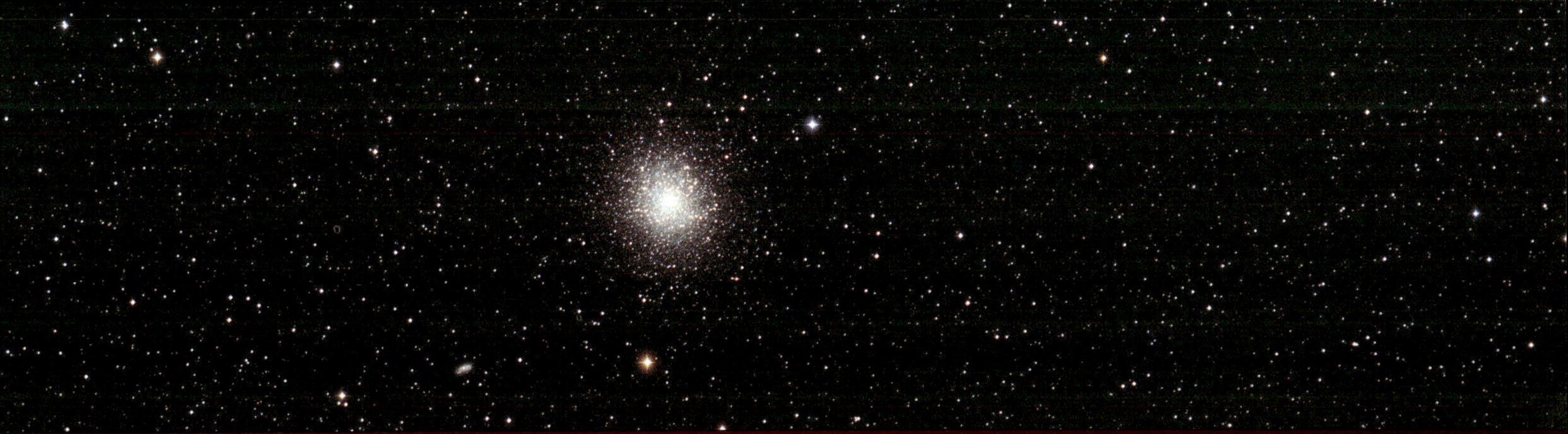

M13 – Globular Cluster

Globular clusters are generally much older than open clusters, and they orbit the galaxy center in a halo off the flat galactic plane of rotation. These clusters contain from tens of thousands to millions of stars in a spherical or globular shape.

M13, the Great Hercules Cluster shown here at three levels of magnification, is estimated at 12 billion years of age, dating almost back to the dawn of the universe. The most visible stars are red giants, with blue stragglers also making a bright showing.

At 25,000 light-years distant in the constellation Hercules and at a diameter of about 150 light-years, this star cluster is one of the largest and brightest globulars visible to earthbound northern hemisphere observers. It is thought to contain about a million stars; at the center the stars are 500 times more concentrated than in our own local solar neighborhood. Night skies must be spectacular there!

M13 wide field (above): 3 hours 20 minutes exposure, SXVF-M25C / Sky-90; medium magnification image at left: 1 hour 34 minutes RGB exposure, SXV-H9C / Hyperstar; and the shallower and more magnified core image below: 15 minutes exposure, Canon EOS-20D / LX200.

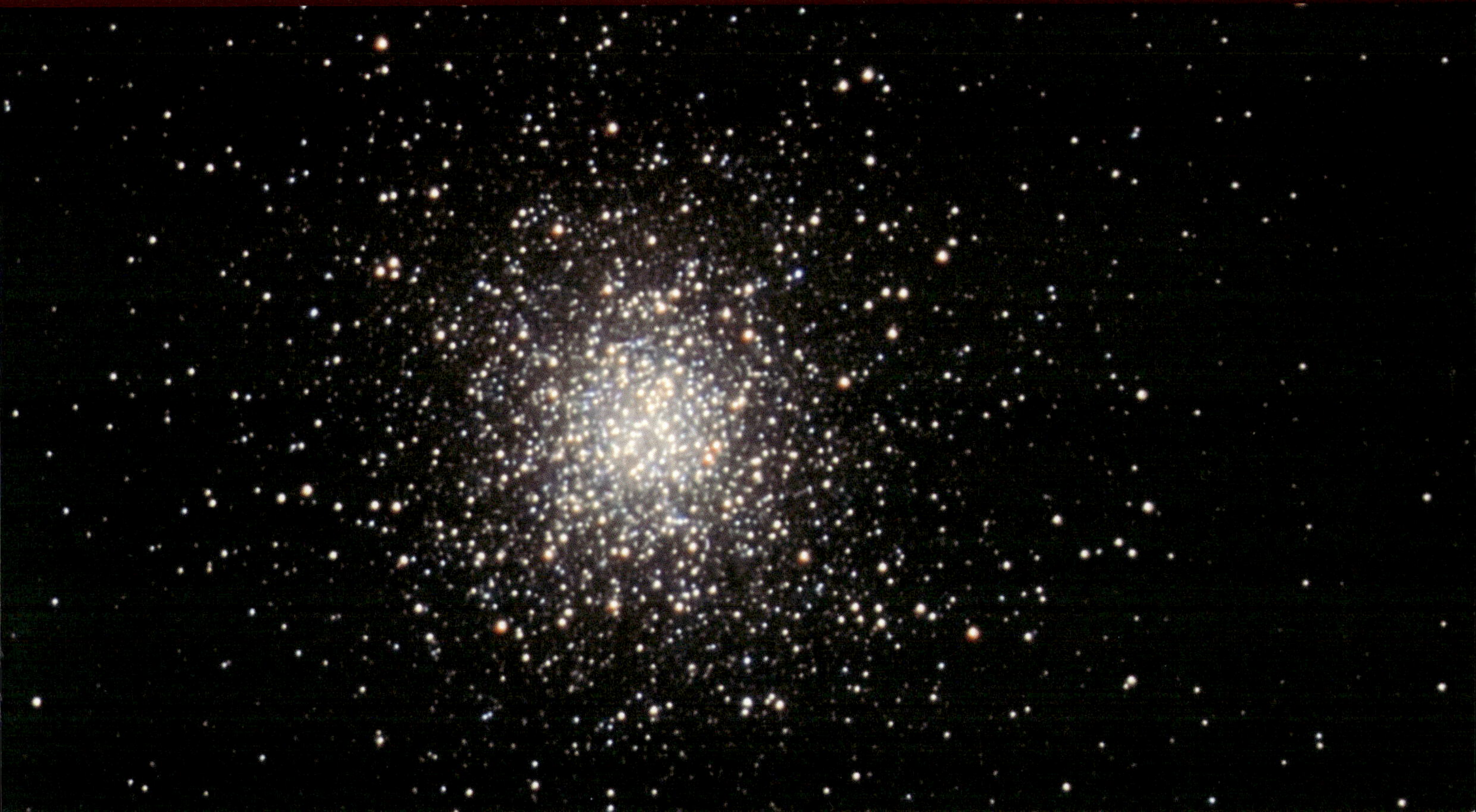

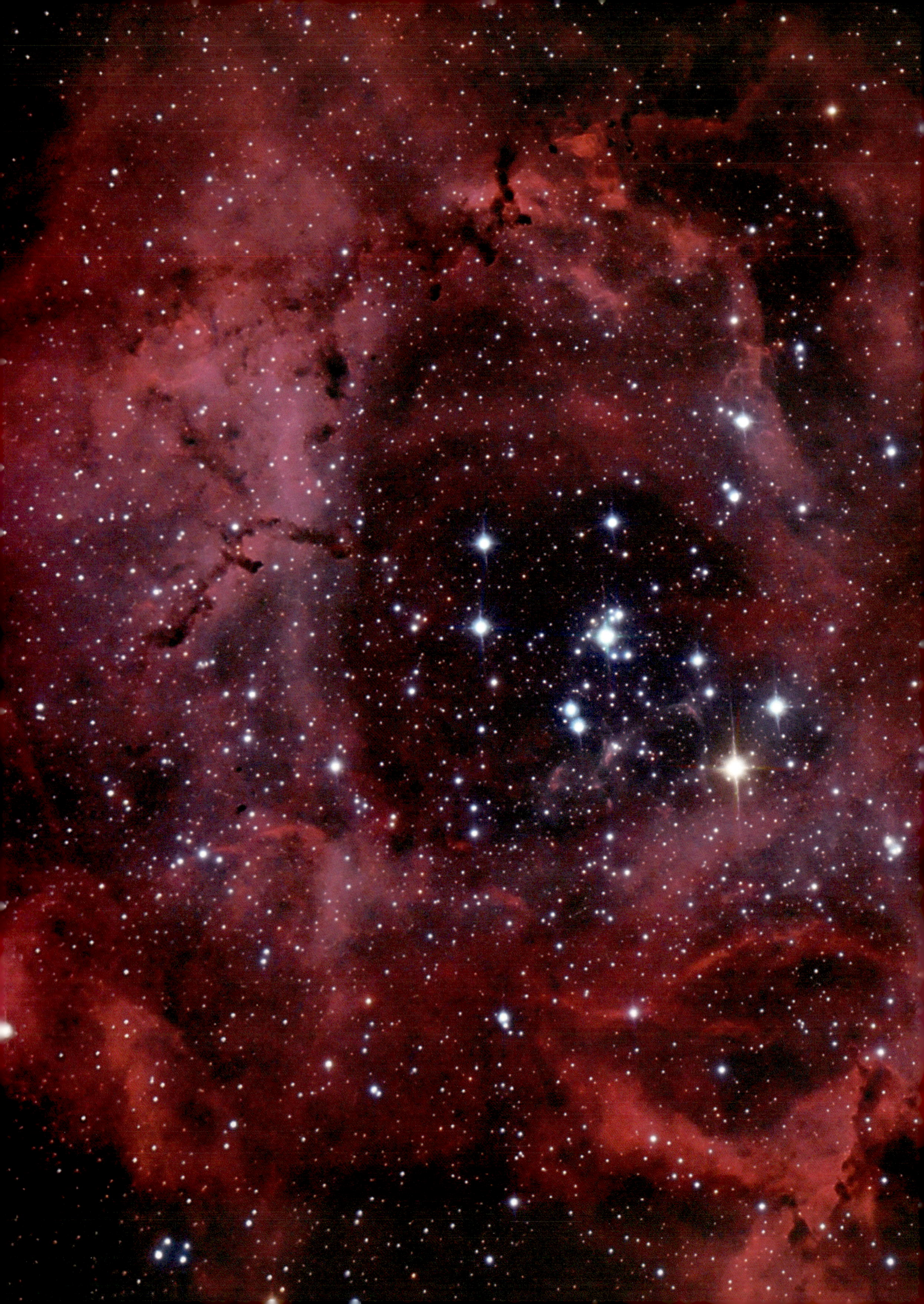

NGC2244 – Star Cluster and Rosette Nebula

Powerful stars sculpt the space around them with their stellar winds, made up of atomic particles blown off the stars at high energy levels. The young stars of open cluster NGC2244 - having formed only about 4 million years ago from the nebula's primal materials - are now pushing gas and dust outward to form a cosmic bubble, a dark open space in the center of the Rosette Nebula.

In the constellation Monoceros, and named for the obvious resemblance to the petals of an open red rose, the Rosette Nebula contains many dark dust clouds - some known as Bok globules, the birthplaces of new stars!

The entire nebula, which extends well beyond the edges of this image (see the next page), spans almost 2 degrees in the sky (a full Moon only covers half a degree)! The Rosette lies at a distance of about 4,900 light-years from us, which is three times further away than the M42 Nebula in Orion, and yet they are of a similar apparent size in the sky. That means the Rosette Nebula spans something like 115 light-years, a truly huge collection of stars, gas, and dust!

Some people say they've seen what resemble a leaping cat, a dog, a duck, a donkey, an elephant, and even a fish in the dark clouds of the Rosette Nebula. Can you find these?

Rosette close-up image: 1 hour 24 minutes RGB exposure, SXV-H9C / Hyperstar.

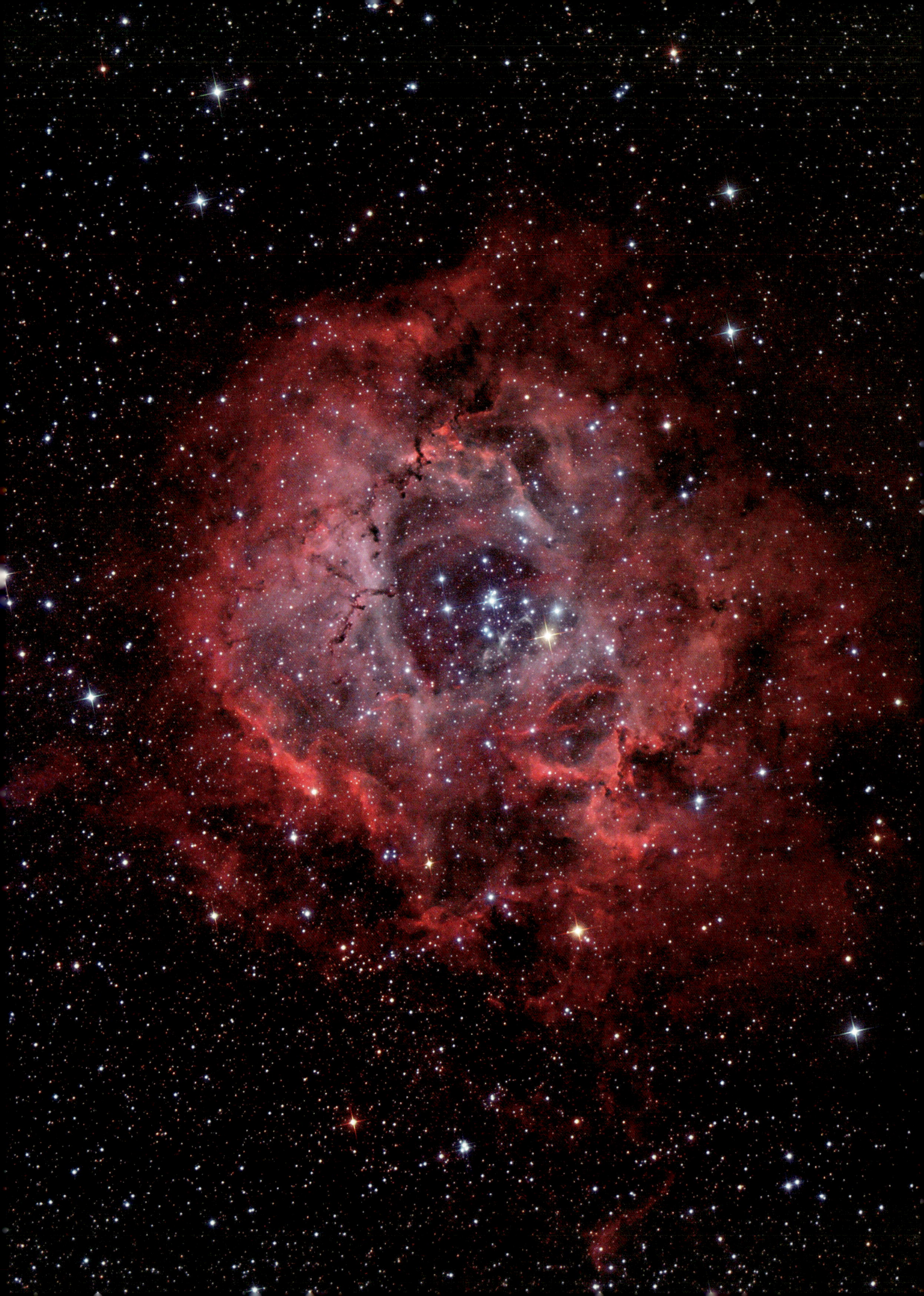

Rosette Nebula Wide Field

A wider and deeper view than the Rosette close-up on the previous page, here we can see that glowing hydrogen gas extends out from the Rosette for many light-years.

There is often confusion over the designation of this nebula complex in the New General Catalog, possibly because the star cluster has a separate designation, NGC2244, and the brighter parts of the nebula each have their own entries, including NGC2237, NGC2238, NGC2239, and NGC2246. To make matters even more confusing, open star cluster NGC2252 is also in the area, at the lower-left edge of the Rosette at about 7 o'clock.

The SXVF-M25C / Sky-90 camera-telescope combination gives a larger field of view than the SXV-H9C / Hyperstar / C11 setup, and captures the whole of the beautiful Rosette nebula in a single frame. Total: 4 hours RGB exposure plus 5 hours 30 minutes exposure through hydrogen alpha emission (Ha - deep red) narrowband filter and 3 hours through oxygen III emission (OIII - teal) narrowband-filter, SXVF-M25C / Sky-90.

M33 – The Triangulum Galaxy

Hot, young star clusters and glowing gas clouds can be seen quite clearly in galaxies nearby but outside our own.

The Triangulum constellation lies near Andromeda and contains a spiral galaxy almost as impressive as the Great Andromeda Galaxy, M31, though it is less well known. A member of our Local Group of galaxies, of which our own Milky Way is also a member, M33 is only about 3 million light-years distant, which is actually quite close to us in intergalactic terms!

In this image we can easily see several very prominent glowing gas clouds (ionized hydrogen emitting light at 656.3 nanometers wavelength in the deep red end of the visible spectrum). The red nebulous region at the upper-left near the orange stars is so prominent that it has its own designation, NGC604, and is very large (about 1,500 light-years across). NGC604 emits a spectrum of light much like our own Great Orion Nebula.

The Triangulum Galaxy occupies an area in the sky almost 4 times as large as the full Moon! Although it appears bright in this long exposure image, the surface brightness of this galaxy is actually quite low, making it a challenging target with a small telescope.

M33 wide field (below): 2 hours 50 minutes RGB exposure, SXVF-M25C / Sky-90; close-up image (right): 2 hours 10 minutes RGB exposure, SXV-H9C / Hyperstar.

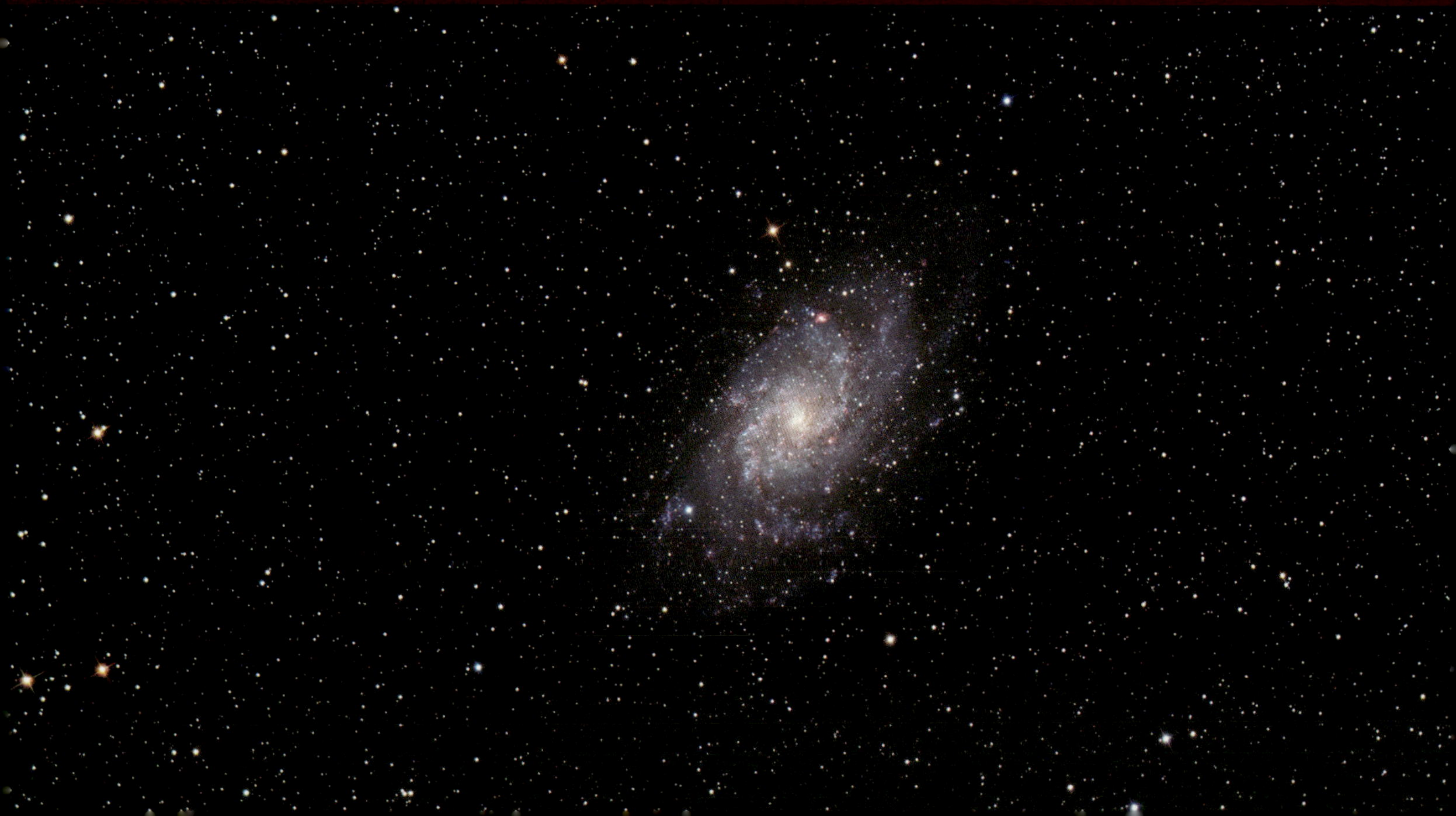

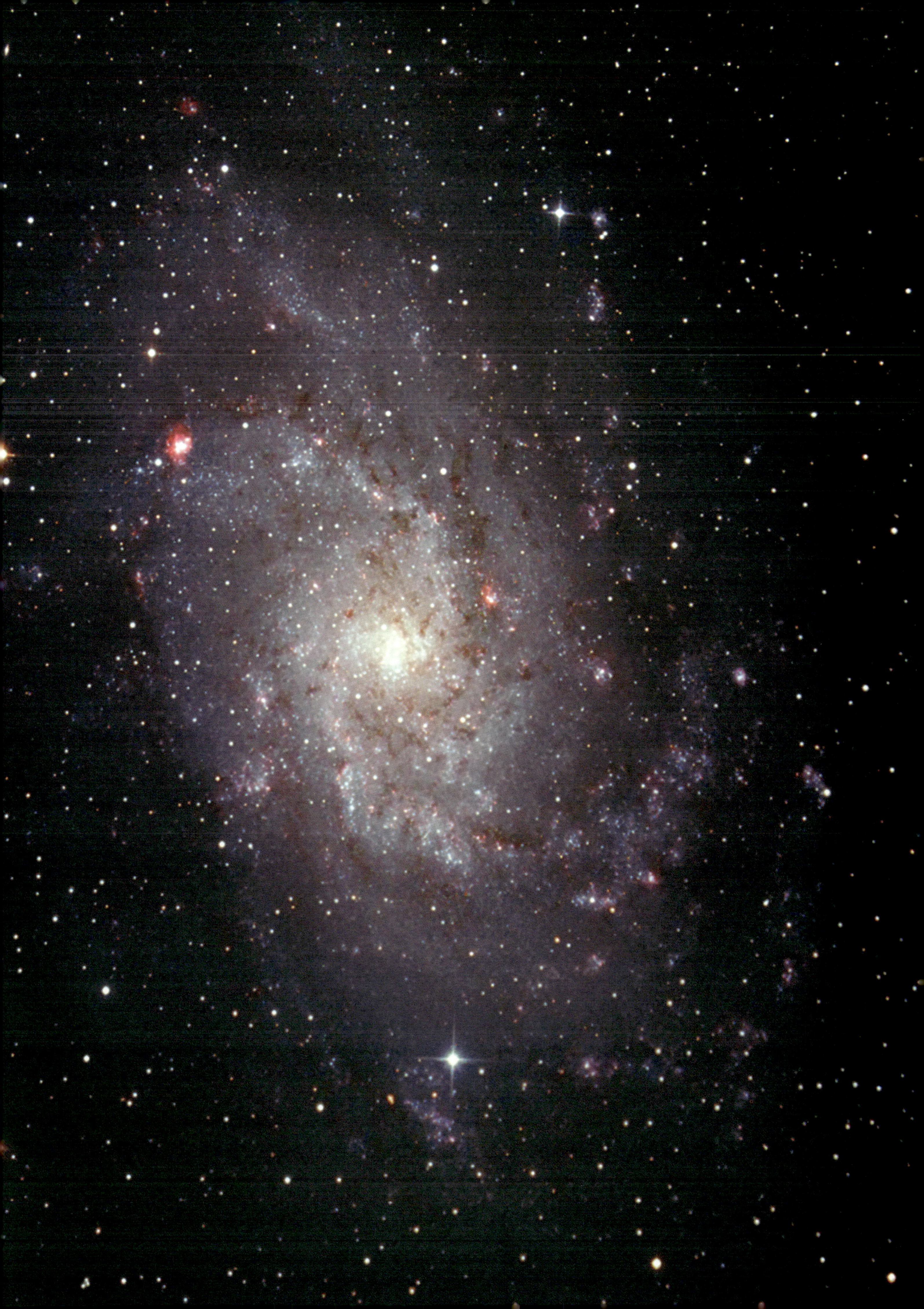

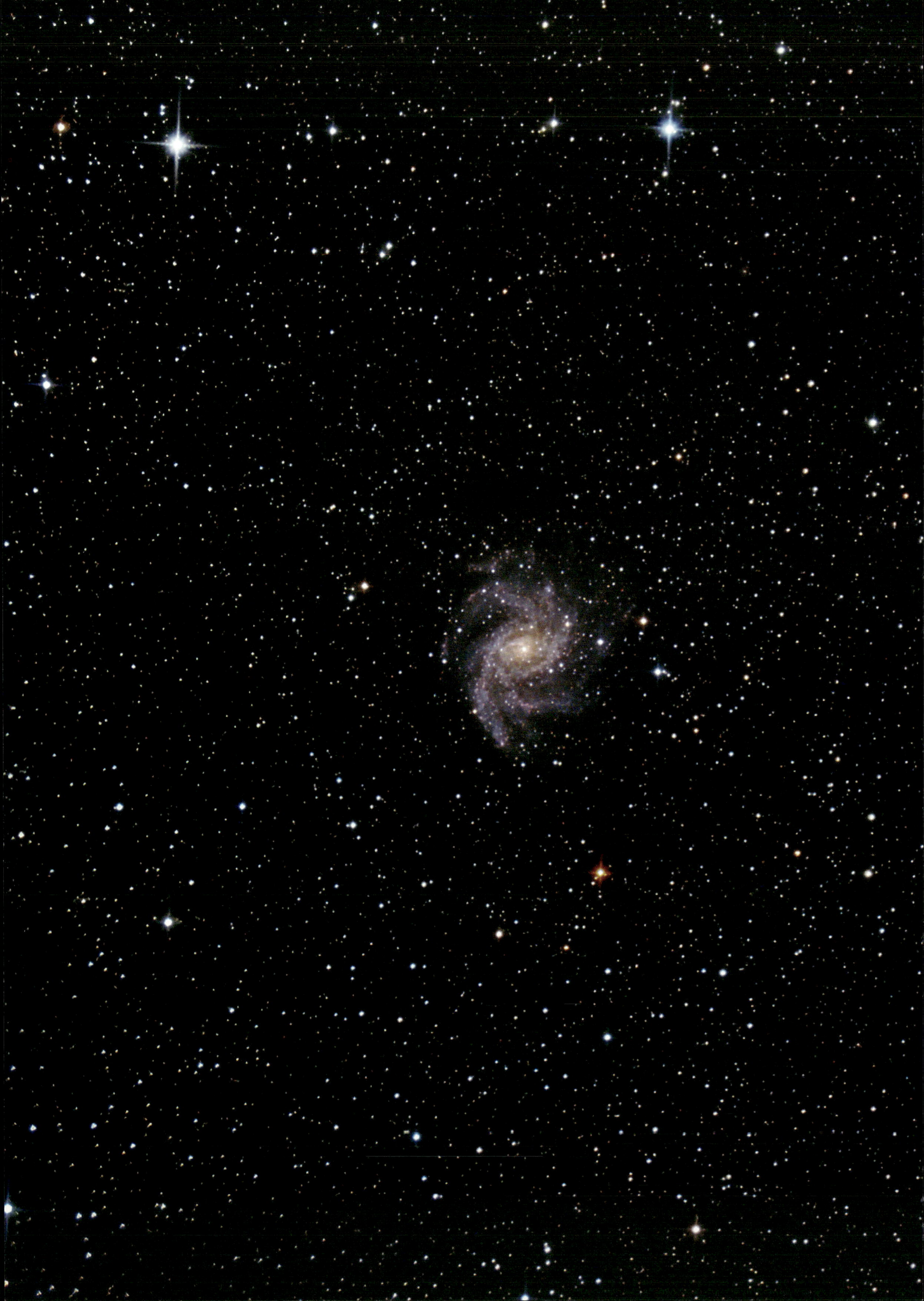

NGC6946 – Spiral Galaxy

Galaxies and stars come in all colors. We can tell a lot about individual stars by their colors and spectra, and the collective color of billions of stars in distant galaxies can tell us about galactic age, phases of star formation, recent interactions with other galaxies, and more. Visually, these celestial wonders paint beautiful pictures indeed!

NGC6946, also known as Caldwell 12, is not far from us on an intergalactic scale (about 15 million light-years distant). Some think it may be an outlying member of our Local Group of galaxies.

We're viewing NGC6946 face-on, and in the constellation Cepheus it's quite close to our galactic plane, which explains the large number of stars from our galaxy also showing in this image. Interstellar gas and dust in the Milky Way actually obscure a lot of the light from NGC6946.

One very interesting thing about this galaxy is that an unusually large number of supernovae have been observed in it over the last 100 years – a total of nine, the most recent seen in 2008. That's about nine times as many violently exploding stars as are typically seen in the average galaxy.

NGC6946 image: 1 hour 10 minutes exposure, SXV-H9C / Hyperstar.

Polaris – The North Star

There happens to be a very bright star visible to the naked eye that is very near the center of rotation of all the objects in the northern sky. This is our pole star, Polaris, often also called the North Star.

Currently, Polaris is almost perfectly aligned with the rotational axis of Earth. Put another way, it's almost straight up from the North Pole, and no matter what the season or where you are it is always north from your location on Earth, and thus it is useful in celestial alignment and navigation.

Believe it or not, Polaris won't always be the North Star, but it will be a long time in human terms - thousands of years - before the precession of the equinoxes moves the axis of Earth to align with other positions in the sky. For example, in about 14,000 years Vega will become the pole star due to this "wobbling" of Earth, like a spinning top, on its axis. What will astronomy be like then? Will our children's children be able to see the stars from their own personal spacecraft?

Polaris image: 1 hour exposure, SXV-H9C / Hyperstar.

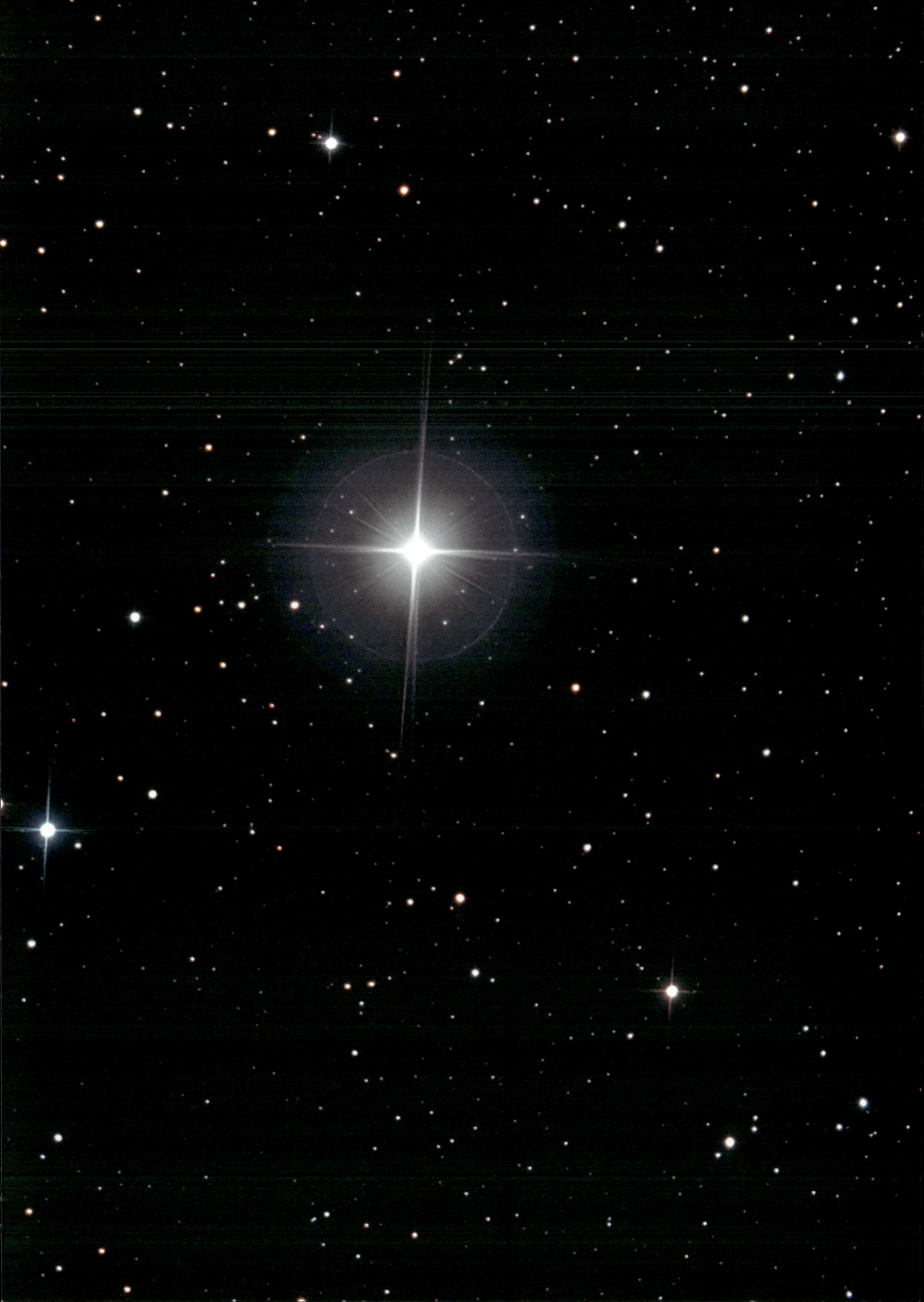

Polaris Is Actually a Multiple Star System!

Some stars have companion stars orbiting them. Polaris is actually a ternary star system, a main star that has two orbiting stellar companions, and it is thus quite interesting astronomically.

Polaris A, the bright star we can easily see, is a supergiant type F7 yellow Delta Cephei variable star that actually pulsates slightly between magnitudes 1.92 and 2.07. In this more highly magnified and less deeply exposed view we can see that it has a less brilliant magnitude 8 companion star called Polaris B, just at the edge of the brighter star's optical flare in this image. Polaris B is in actuality almost 190 billion miles from Polaris A, 12 light-days distant, and at such a large distance Polaris B takes thousands of years to complete a single orbit.

Polaris A also has a much closer companion as well, not visible in this image. It's called Polaris Ab and is known to exist thanks to careful light level measurements by astronomers over the past 50 years. In fact, it is beyond the resolution capabilities of virtually all ground-based telescopes today and has only recently been directly imaged for the first time by the Hubble Space Telescope. A little more massive than our own Sun, Polaris Ab is 2 billion miles from Polaris A, about as far out as the orbit of the planet Uranus from our own Sun.

The Polaris multiple-star system lies about 400 light-years from Earth in the constellation Ursa Major.

Polaris high resolution image: 12 minutes exposure, Canon EOS-20D / LX200.

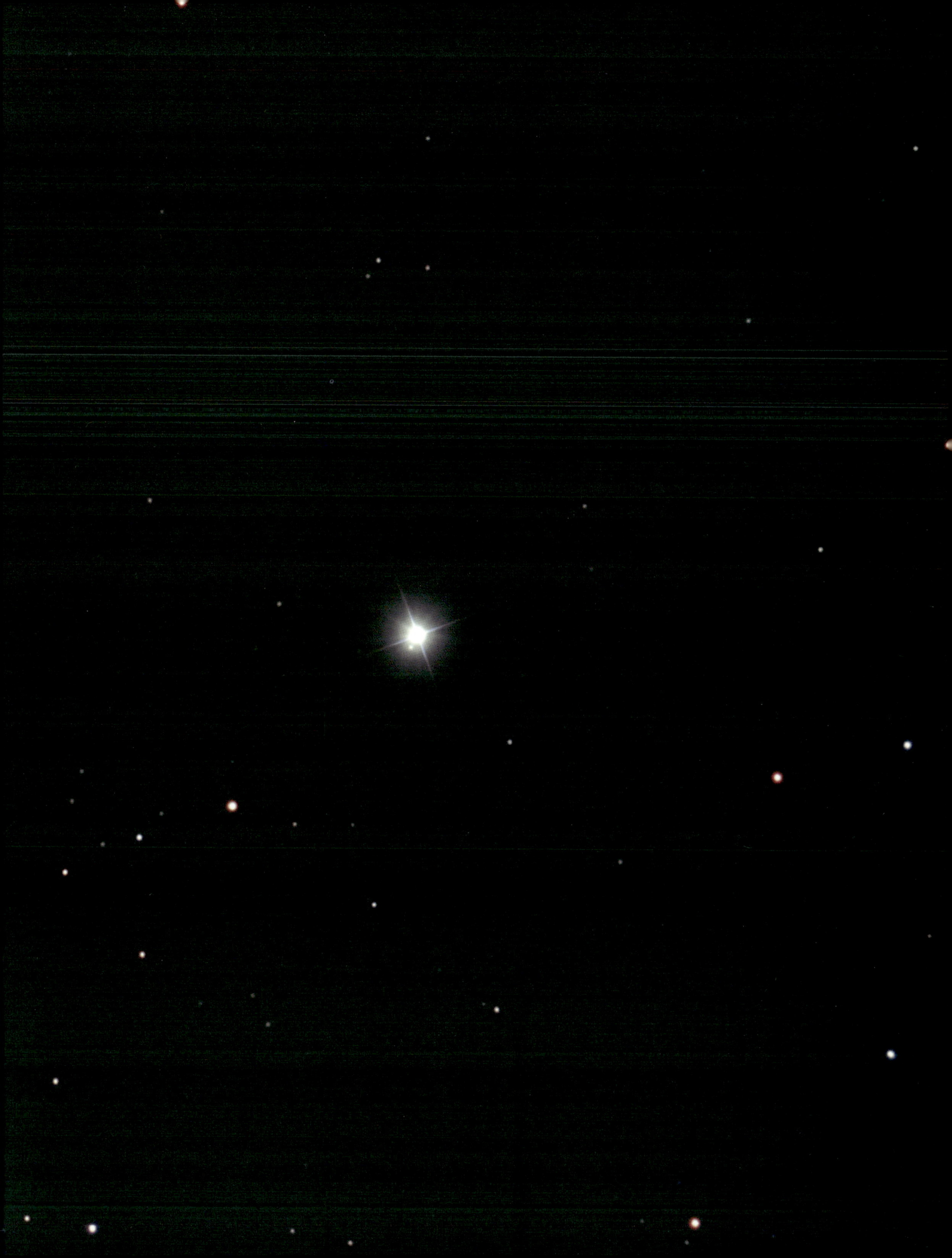

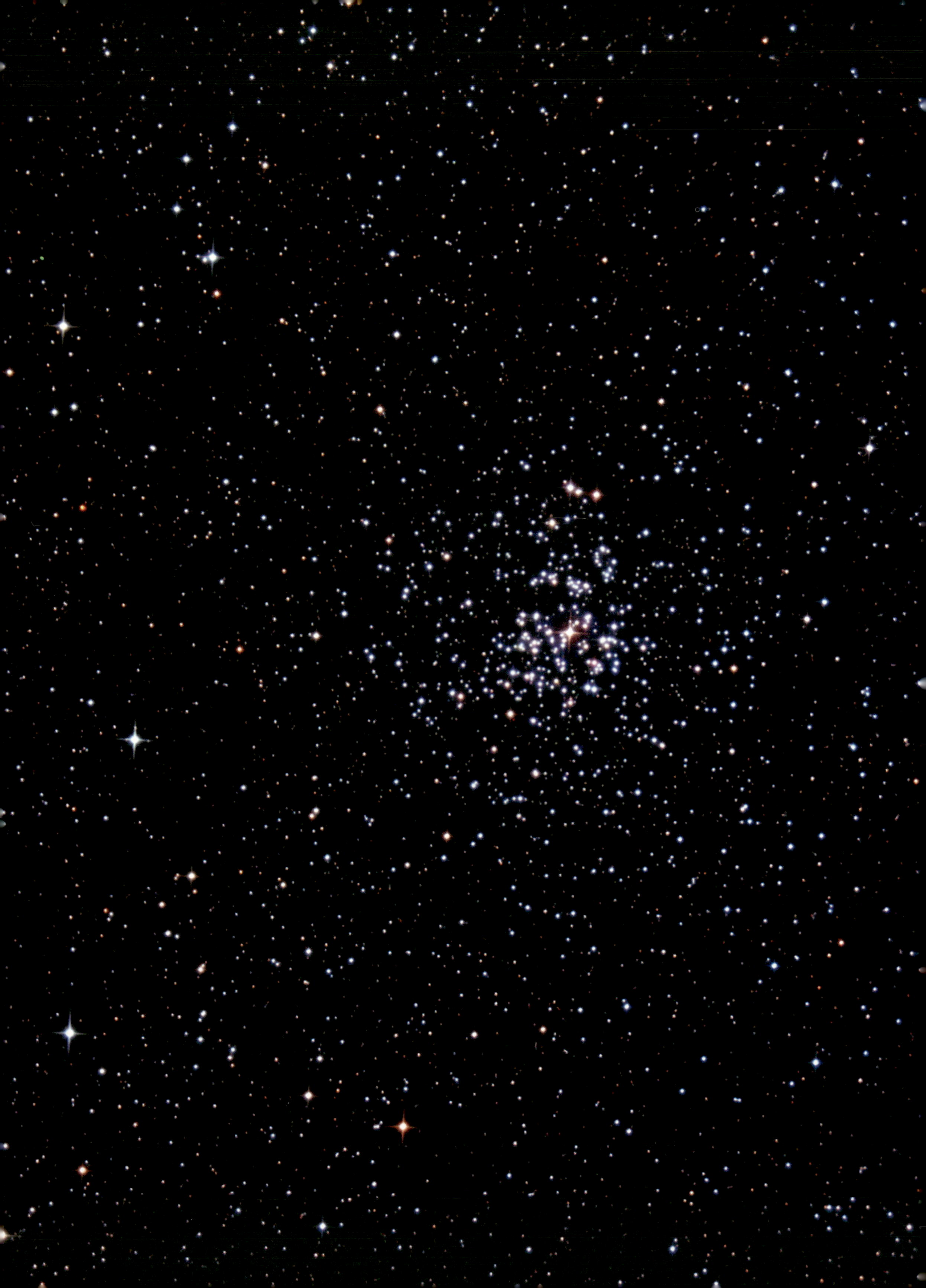

M37 – Open Cluster

Though we often use the Messier catalog from the mid-1700s to refer to the brighter objects in the sky, this particular cluster was seen and cataloged much earlier – in 1654 in Palermo, Italy, by Giovanni Batista Hodierna. Hodierna used just a small Galilean refractor with a magnification of 20. He described this cluster as *Nebulosae*, which means "nebulous" to the naked eye, but resolved into stars in a telescope.

The richest and brightest of three open clusters in the southern part of the constellation Auriga, Messier object 37 is about 4,000 light-years distant and contains roughly 500 stars across a space of about 25 light-years. 150 stars are bright enough to be seen in a small telescope.

M37 image: 30 minutes exposure, SXV-H9C / Hyperstar.

NGC2175 – The Monkey Head Nebula

Many objects in the sky are named for Earth-bound objects they bring to mind. Can you see the resemblance to a monkey's head here? (Hint: Turn the book 90 degrees clockwise.)

Often mistakenly called NGC2174 (which is a smaller nebulosity within the Monkey Head), this is a region of strong hydrogen-alpha emission in the Orion constellation.

There is some evidence this nebula was seen and noted as early as 1654 by Italian astronomer Giovanni Battista Hodierna, an astronomer at the court of the duke of Montechiaro.

NGC2175 image: 2 hours 37 minutes exposure, SXV-H9C / Hyperstar.

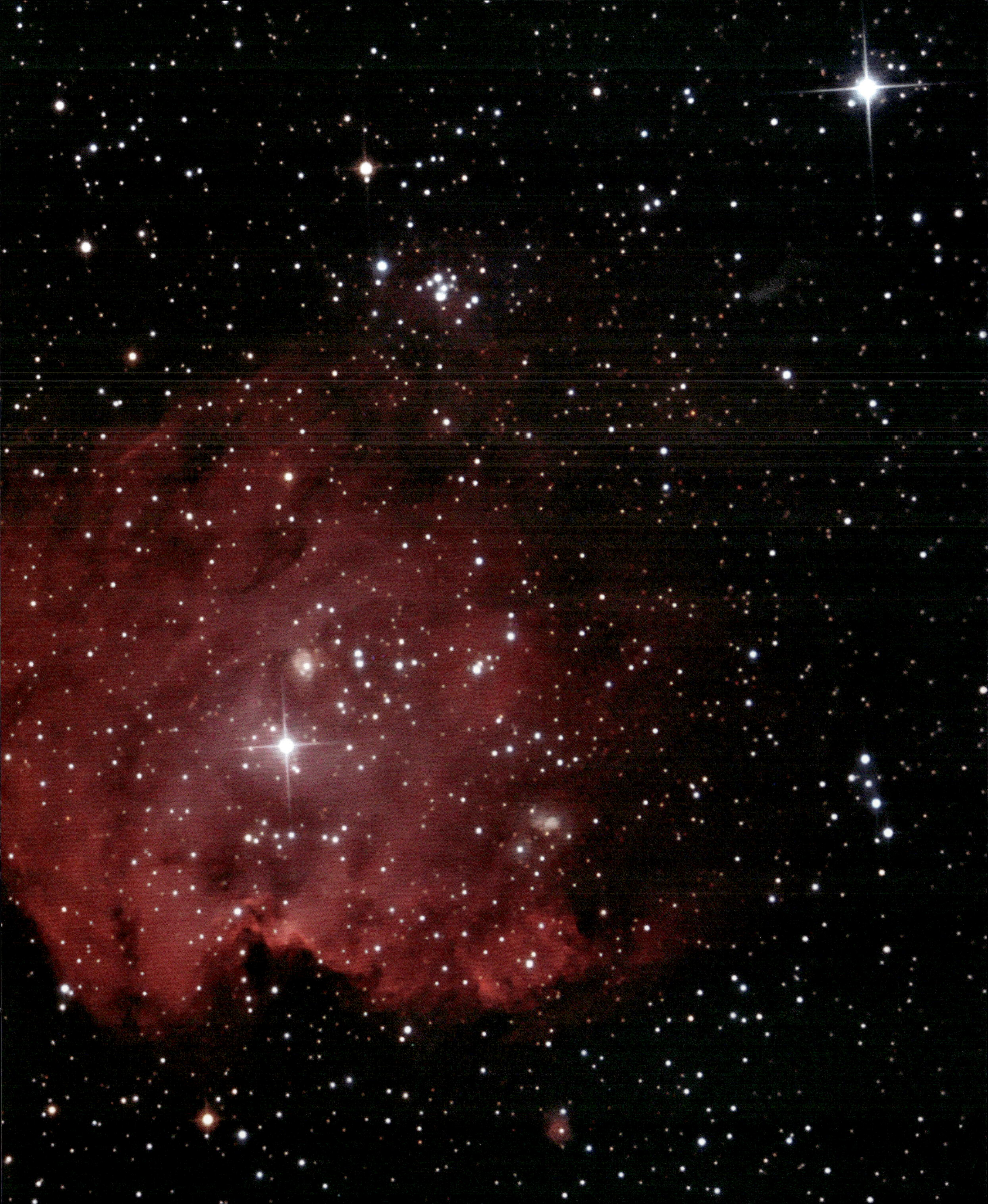

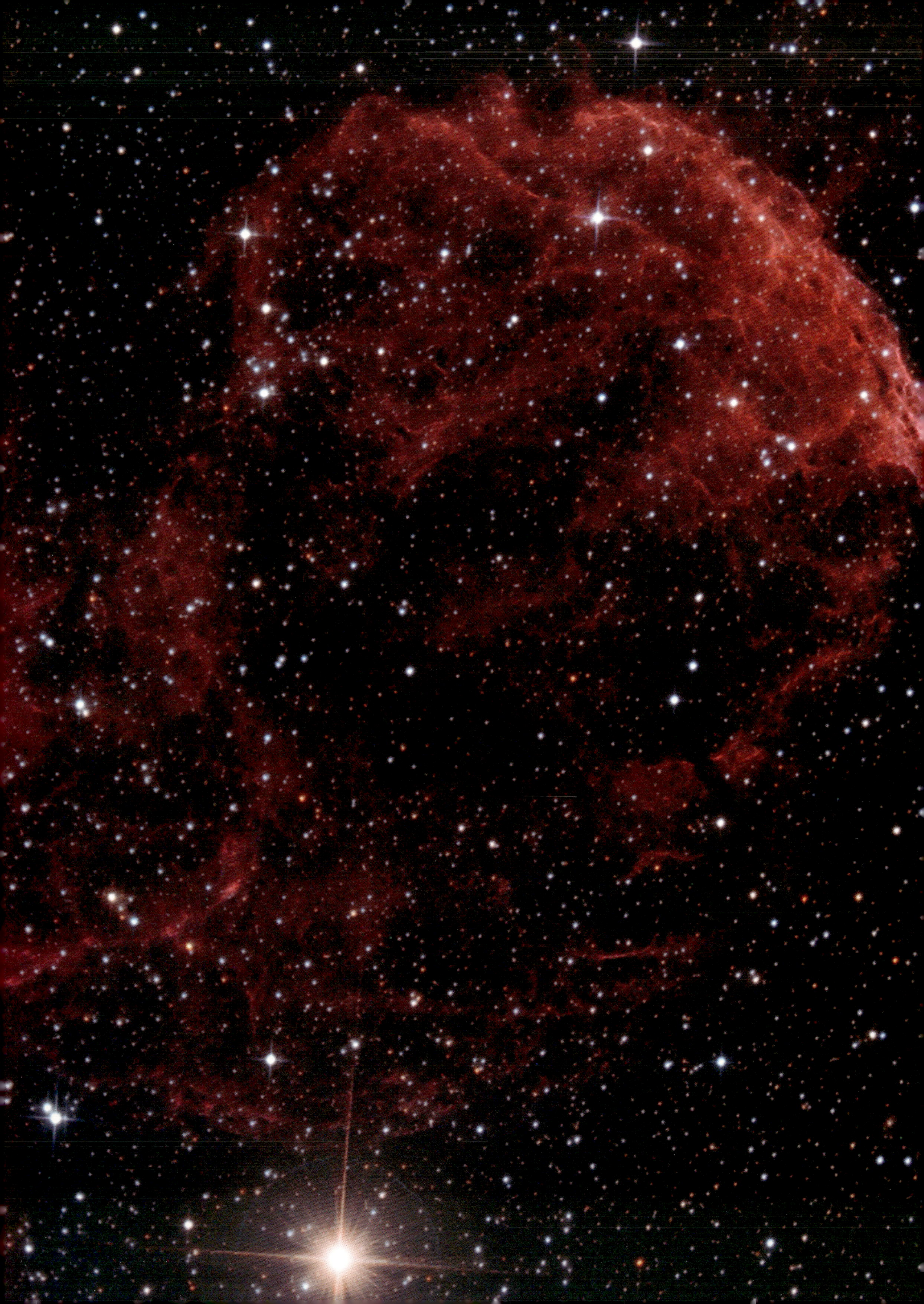

pposite

C443 – The Jellyfish Nebula

faint but large nebula is a remnant of an ancient supernova explosion took place in the constellation Gemini. The Jellyfish Nebula is also known C443 in catalogs.

ut twice in every century in our Milky Way Galaxy a massive star odes violently, becoming a supernova. When the nuclear furnace at the of a particularly large star exhausts its fuel, the core collapses in a few nds into a neutron star or black hole. As the outer parts of the star fall the now tiny remnant of the core, the material becomes superheated to ns of degrees, and an extremely violent explosion occurs, expelling erial outward at incredible speeds.

bout 5,000 light-years distant, the filaments we can see in the Jellyfish ula are the visible effects of the huge blast wave of expanding material the ancient supernova. The giant wave impacts the molecular cloud of ogen surrounding the area, which is heated and ionized, causing it to . Estimates differ as to how long ago the star exploded; these range 8,000 to 30,000 years ago. The tiny core remnant, in the form of a ron star, has been detected with X-ray telescopes. It is not visible in this e, but if it were, it would be near the center-left edge of the page.

fish Nebula image: 5 hours exposure, SXV-H9C / Hyperstar.

Next Two Pages >

IC443 – Wide Field

A bit deeper and much wider view of the Jellyfish Nebula area is presented here, showing additional emissions and reflection nebulae in the area.

IC443 wide field image: 3 hours RGB exposure, 9 hours 40 minutes Ha exposure, 3 hours 40 minutes SII (sulfur) exposure, SXVF-M25C / Sky-90.

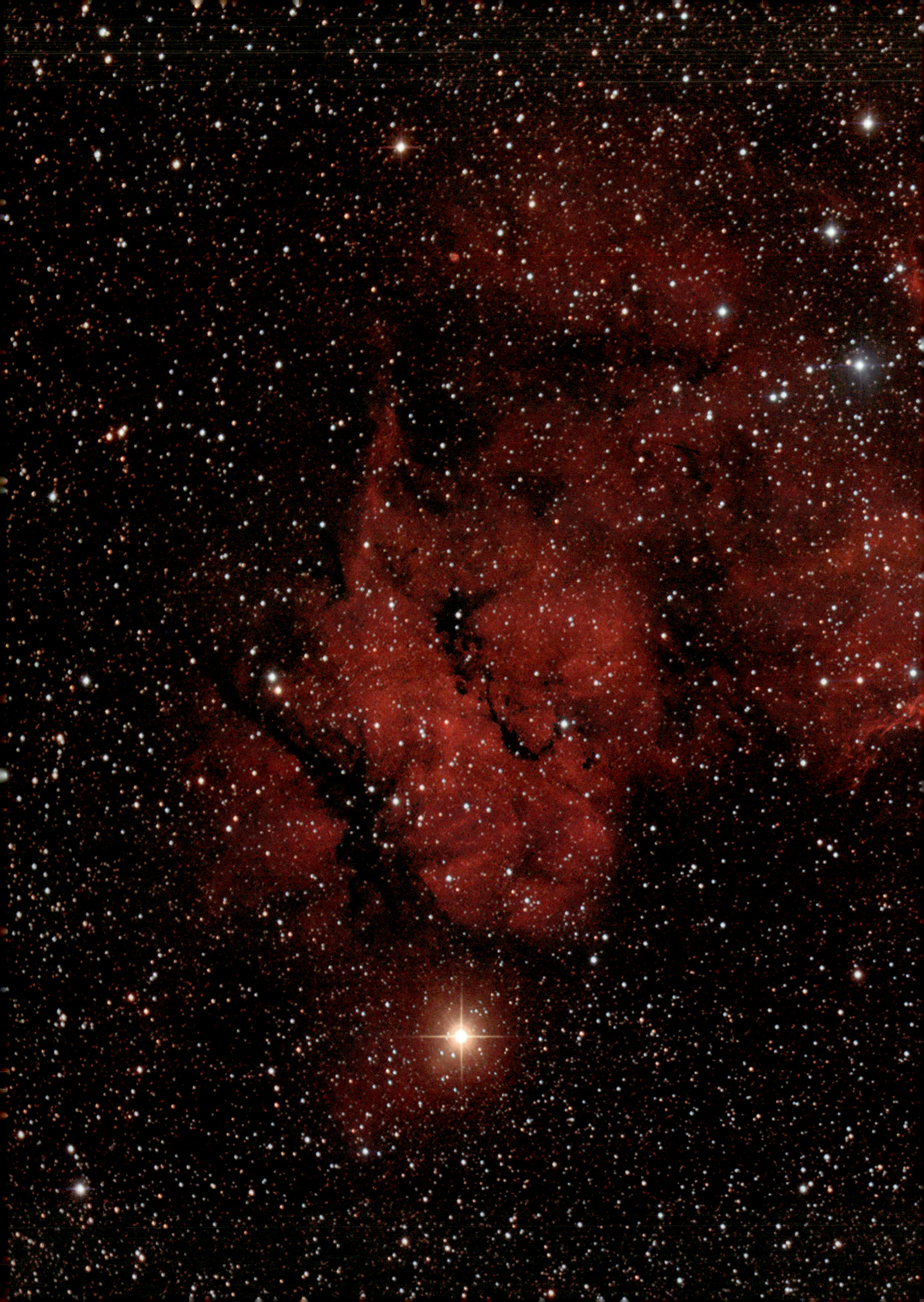

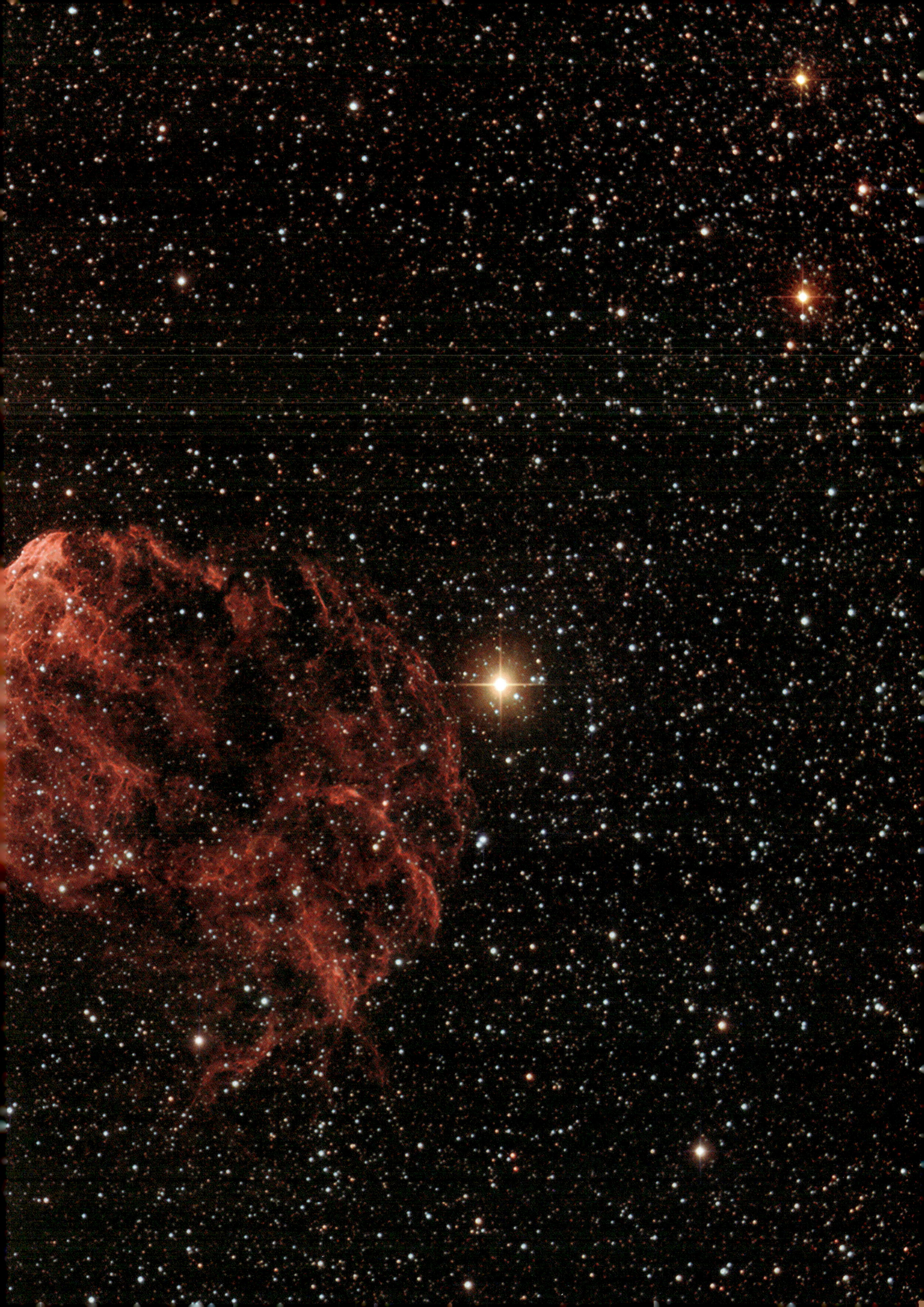

Orion - Very Wide Field

A prominent and familiar sight in winter skies in the northern hemisphere, Orion the Hunter seems to be standing with bow drawn, ready to fire an arrow off to the west. In the sky Orion is standing with his two hunting dogs, Canis Major and Canis Minor, fighting Taurus the Bull.

There are a lot of astronomically interesting areas in Orion, some of which are visible in this relatively shallow image. At the eastern end of the three stars that make up Orion's Belt we find the red Horsehead and Flame nebulae. Nearby we find the reflection nebula M78, and the Sword of Orion, hanging below his belt, contains the Running Man Nebula, M43 and M42 (the Great Orion Nebula), within which is the Trapezium cluster.

Orion image: 10 minutes exposure, Canon EOS-20D / 17-40 zoom lens.

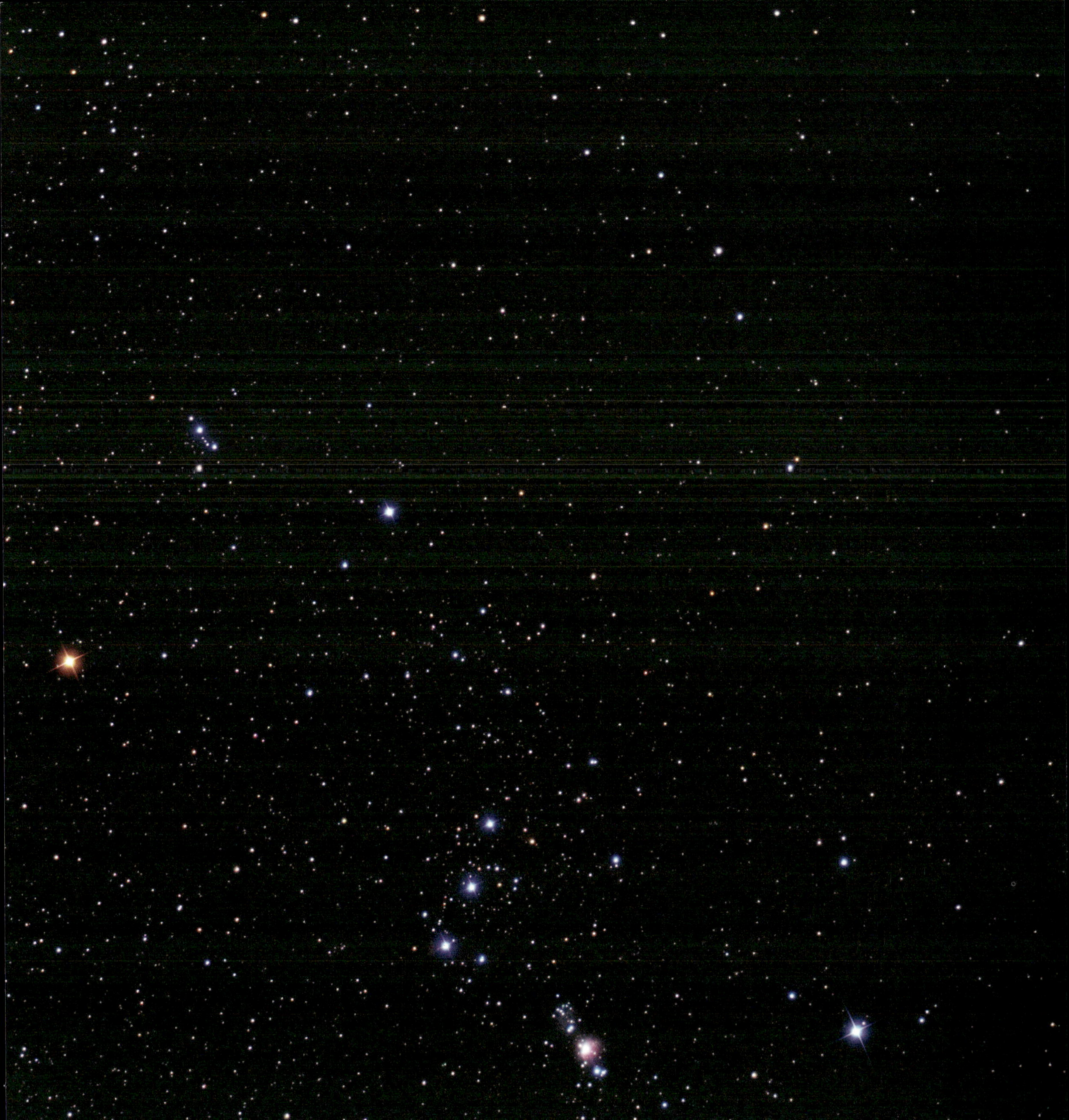

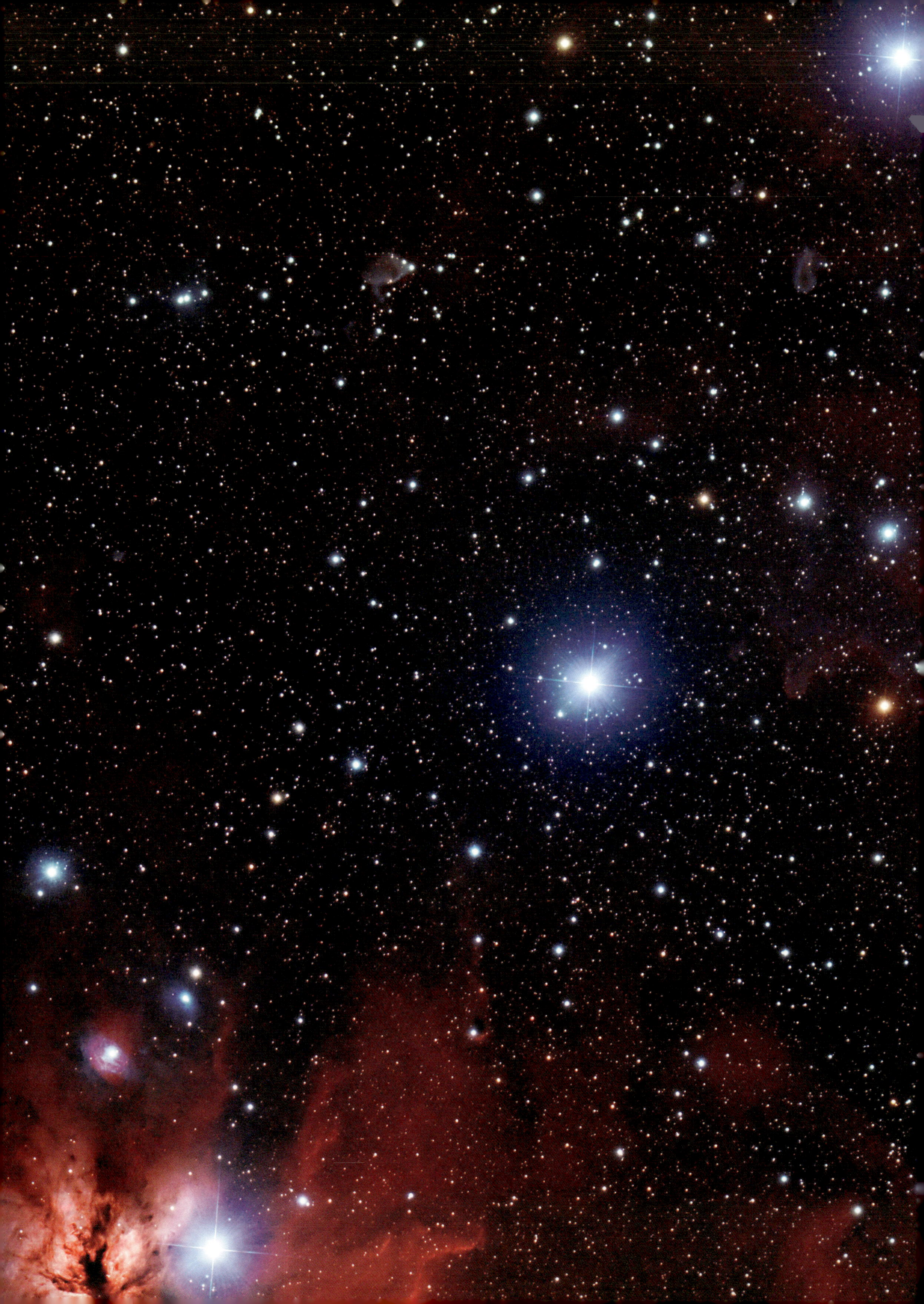

Orion's Belt – Deep Field

A deeper, closer view of Orion's Belt reveals not only more stars but both reflection and emission nebulae throughout. In fact, this is but a small part of the Orion molecular cloud complex, a massive collection of gas and dust in our Milky Way Galaxy that spans the entire constellation.

The nebula at the lower-left corner of this image, next to the bright blue belt star Alnitak, is called the Flame Nebula.

New star formation is currently occurring in the Orion molecular cloud complex. Of special interest are some of the small, bluish, dense-looking reflection nebulae, some of which may be future star nurseries. Can you see the one that looks a bit like a baby's footprint?

Orion's Belt image: 5 hours exposure, SXVF-M25C / Sky-90.

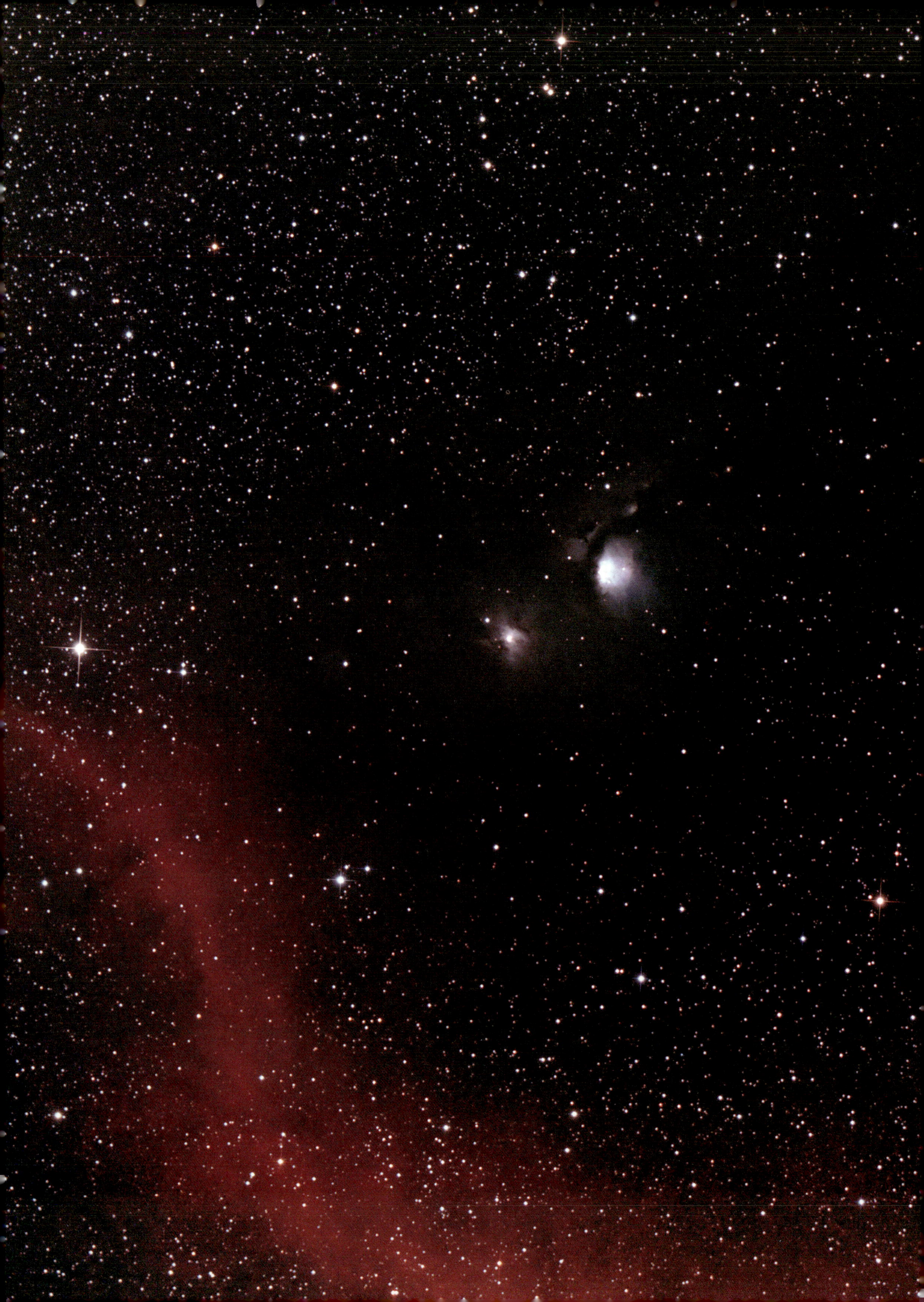

M78 – Reflection Nebula

Our galaxy is rich with monstrous clouds of gas and dust in Orion. M78, a bright reflection nebula, is part of the giant Orion molecular cloud complex.

At left a bit of Barnard's Loop glows red with hydrogen gas, in M78 some of the dust reflects bluish starlight, and dark dust lanes in the foreground become visible in silhouette. In a deep close-up a subtle brownish color in the dust begins to become apparent.

M78 wide-field image (left): 4 hours, SXVF-M25C / Sky-90; close-up image below: 1 hour 54 minutes, SXV-H9C / Hyperstar.

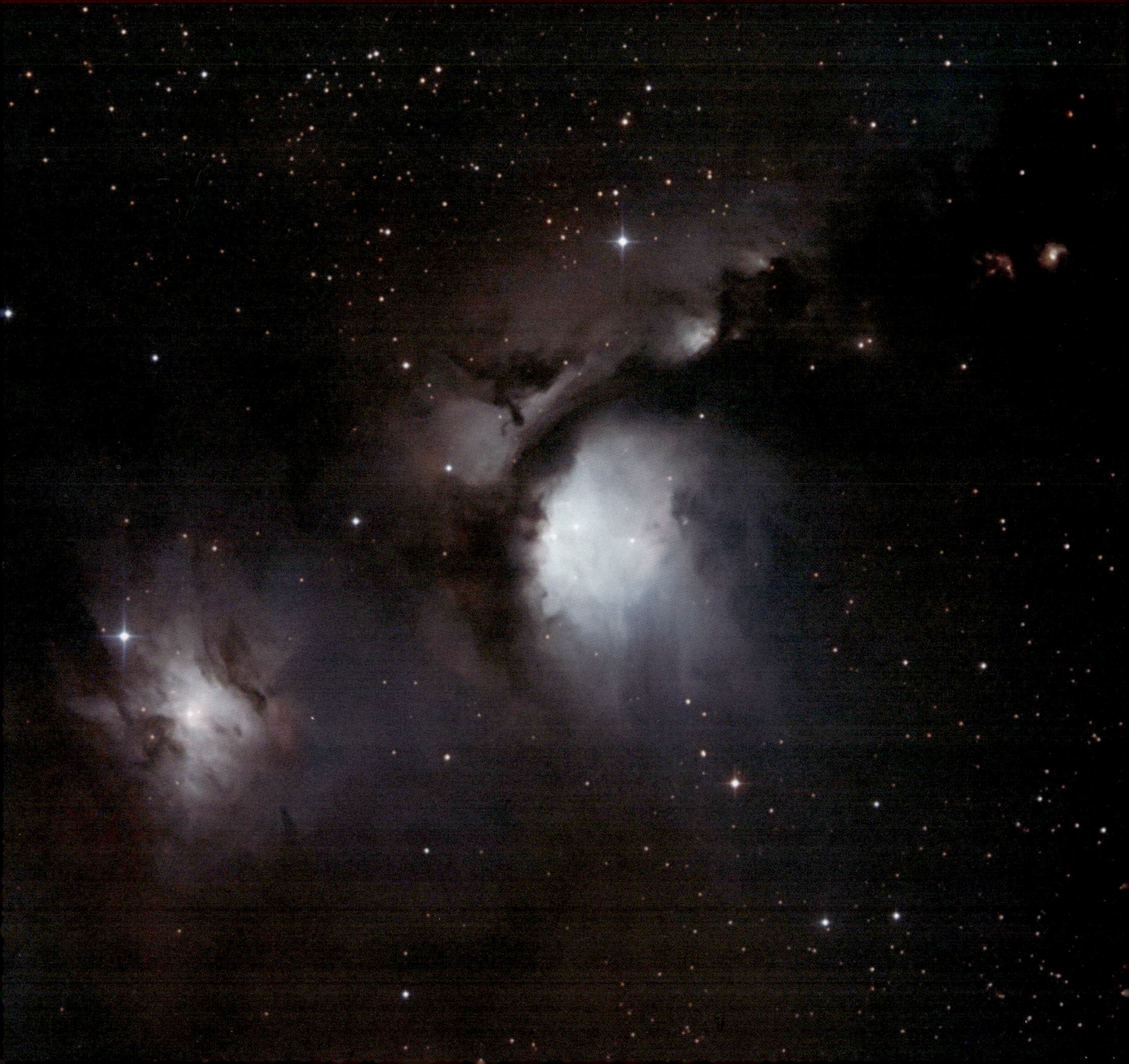

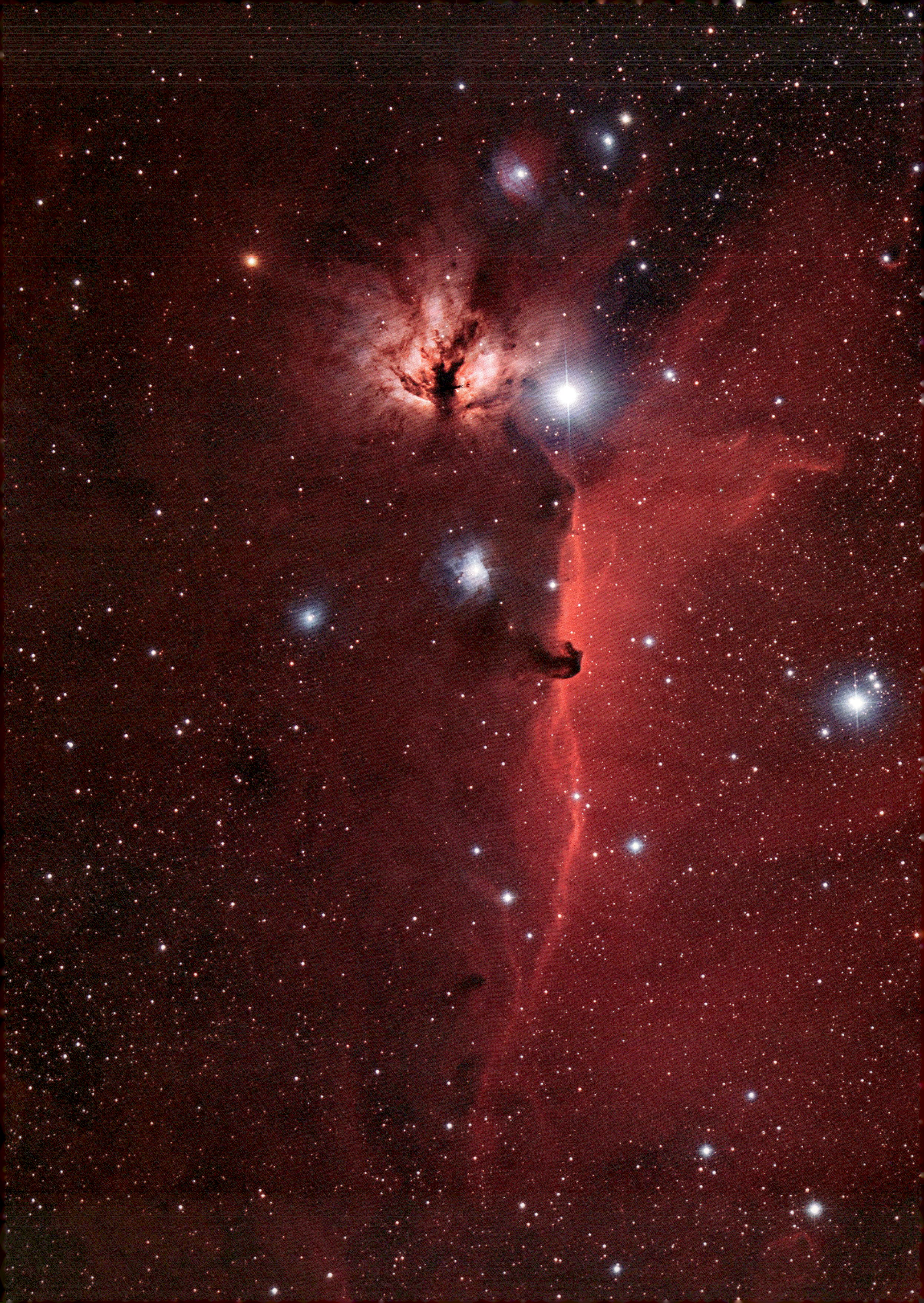

pposite

orsehead Area-Wide Field

r the brilliant eastern belt star Alnitak (above center in this image) in n, a deep image shows glowing hydrogen gas and intervening dark dust comprise one of the most widely recognized nebulae in the sky, the sehead Nebula. Additionally we see the lesser known but still spectacular ne Nebula. There are also several beautiful blue reflection nebulae here.

give you a sense of scale, the size of the horse "head" itself from the tip ne "nose" to "mane" is about 1.5 light-years, which is about 1,000 times diameter of the orbit of Pluto around our Sun. That's one big horse!

sehead area wide-field image: 6 hours RGB and 4 hours Ha exposure, F-M25C / Sky-90.

Next Two Pages >

Flame and Horsehead Close-up

Named for its obvious resemblance to an animal of the equine variety, the dark Horsehead Nebula is actually a star forming region – a swirling cloud of dust that blocks the light from the ionized gas cloud behind it, which in turn has been excited by the light from the energetic and brilliant nearby star Alnitak. The Horsehead Nebula can also be found in catalogs as Barnard 33 or IC434.

The Flame Nebula, so named for obvious reasons, is also a combination of glowing gas and intervening dust and can be found in catalogs as NGC2024. We also see blue reflection nebulae NGC2023 and IC435 in this image.

Flame/Horsehead close-up image stitched from two frames comprising 4 hours 21 minutes total exposure time, SXV-H9C / Hyperstar.

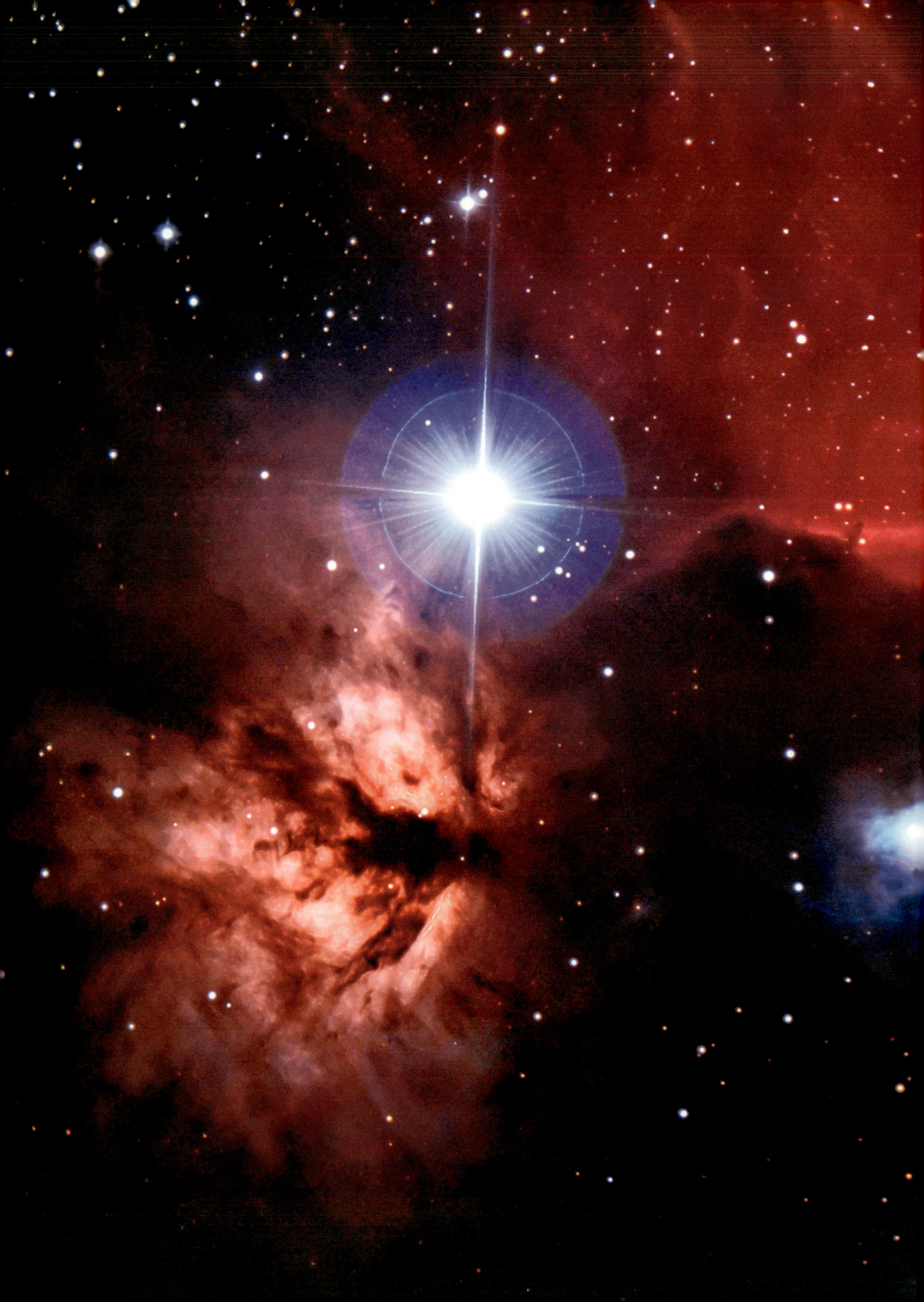

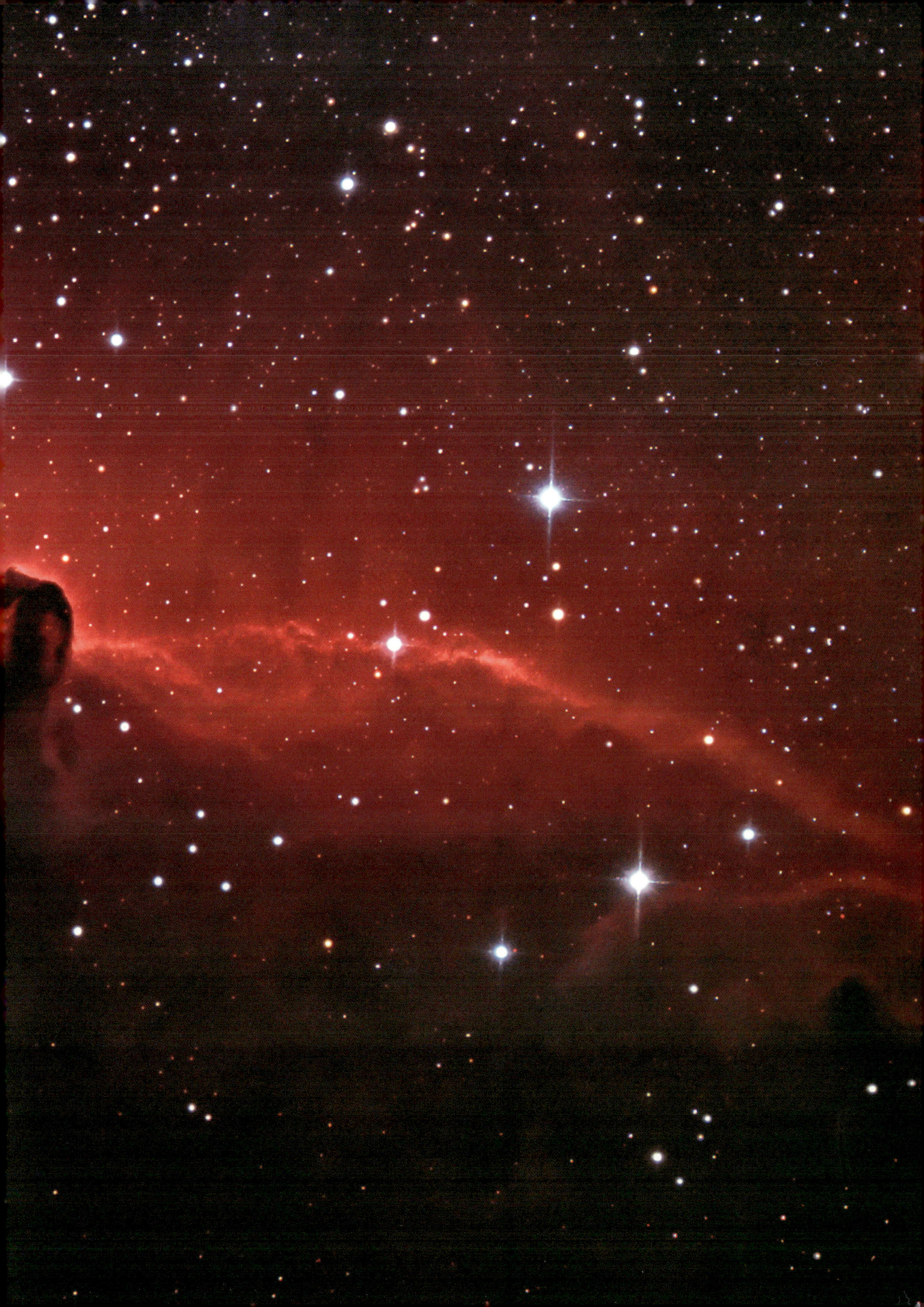

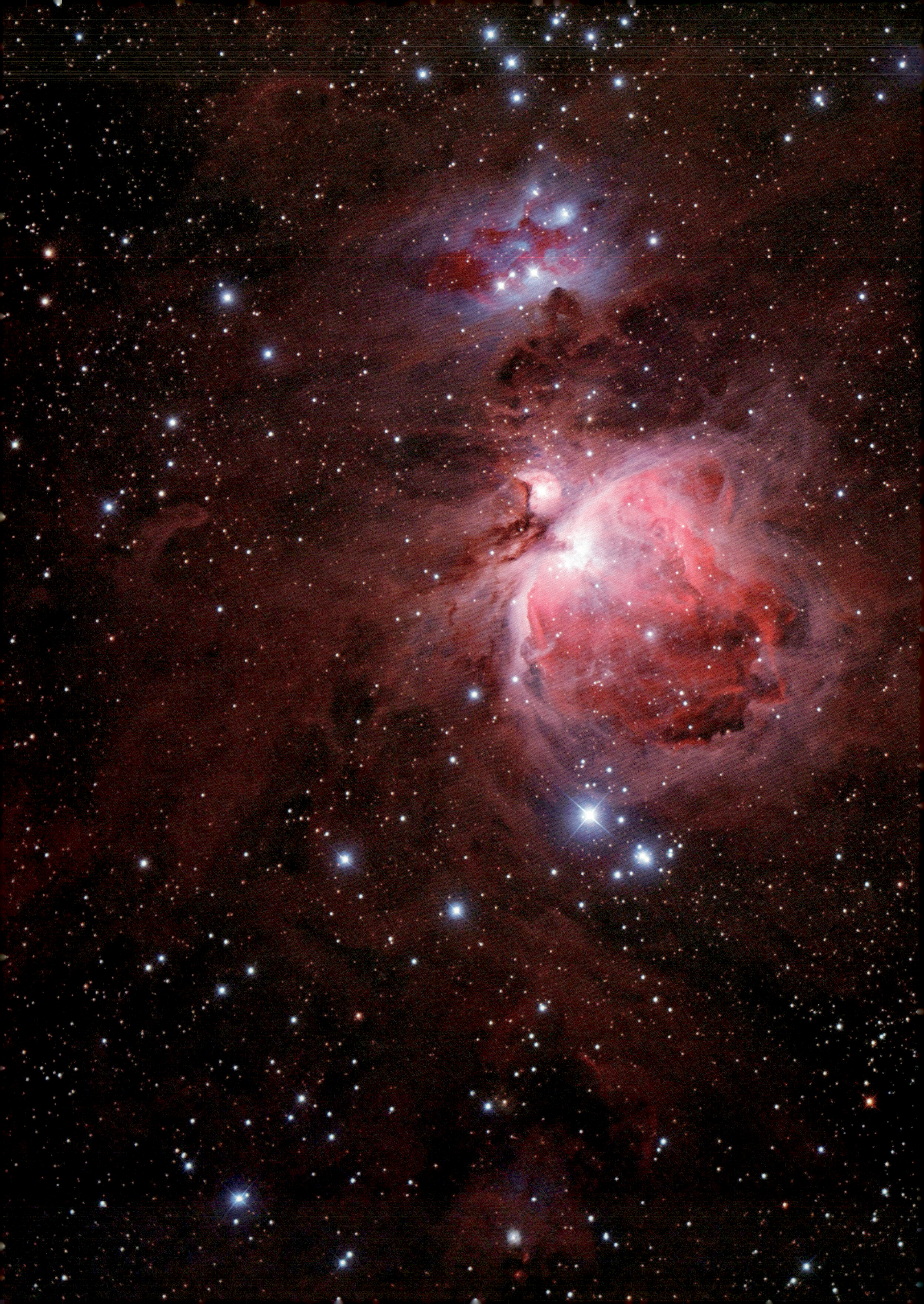

The Sword of Orion

As we move down in Orion to the stars we perceive as Orion's Sword, we see in this deep, wide-field view that it's not just made of bright stars but also quite a spectacular group of nebulae. All varieties are represented here: emission nebulae, reflection nebulae, and intervening dark absorption nebulae.

There are many interesting objects in this image, including the well-known Running Man Nebula near the top, the brilliant M43/M42 complex (sometimes called the Orion Nebula) at the center, and the bright star Hatysa. There are even some dim objects of note toward the bottom, including Herbig-Haro shock fronts that indicate the recent birth of stars; a small rare yellow reflection nebula; and the tiny, bright NGC1999 reflection nebula.

Sword of Orion image: 12 hours total RGB, Ha, and Infrared (R72) exposure time, SXVF-M25C / Sky-90

Interstellar Thunderclouds

Between the Running Man Nebula and the brilliant M43/M42 nebulae, a deep exposure reveals turbulent areas of rich brown dust that some say resemble thunderclouds.

At about 1,600 light-years distant, star formation is actively taking place throughout this area, and it is being studied not only in visible light but in other parts of the spectrum as well. Infrared light, for example, can penetrate dust that blocks visible wavelengths, and thus exposures taken through infrared filters show many more stars here.

This is an active area – the dust is swirling and moving around, and it sometimes obscures more or less of the visible light from stars, causing them to vary in brightness over time. The stars exhibiting this phenomenon have become known as "dust variables."

Thunderclouds image: 1 hour 49 minutes, SXV-H9C / Hyperstar.

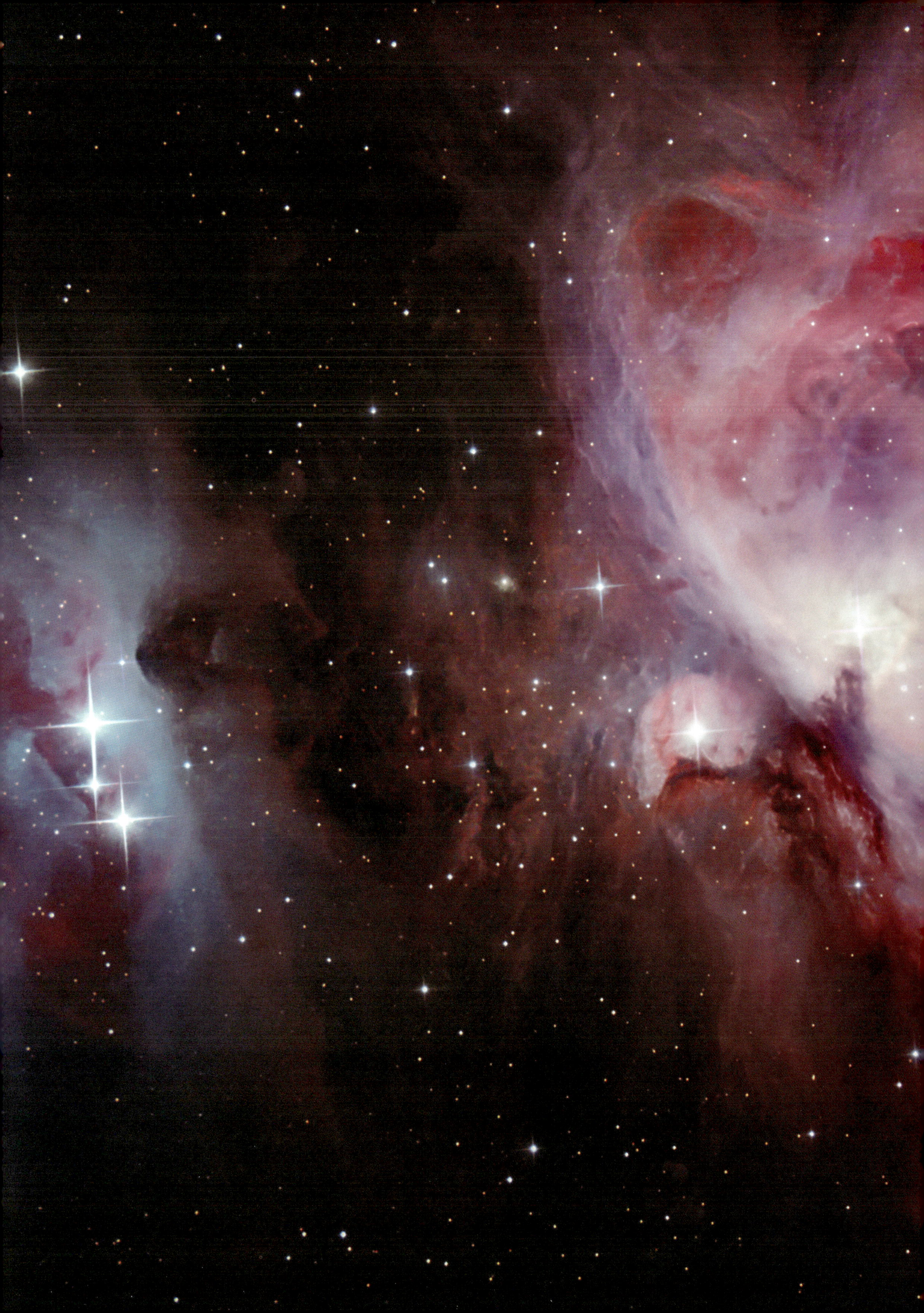

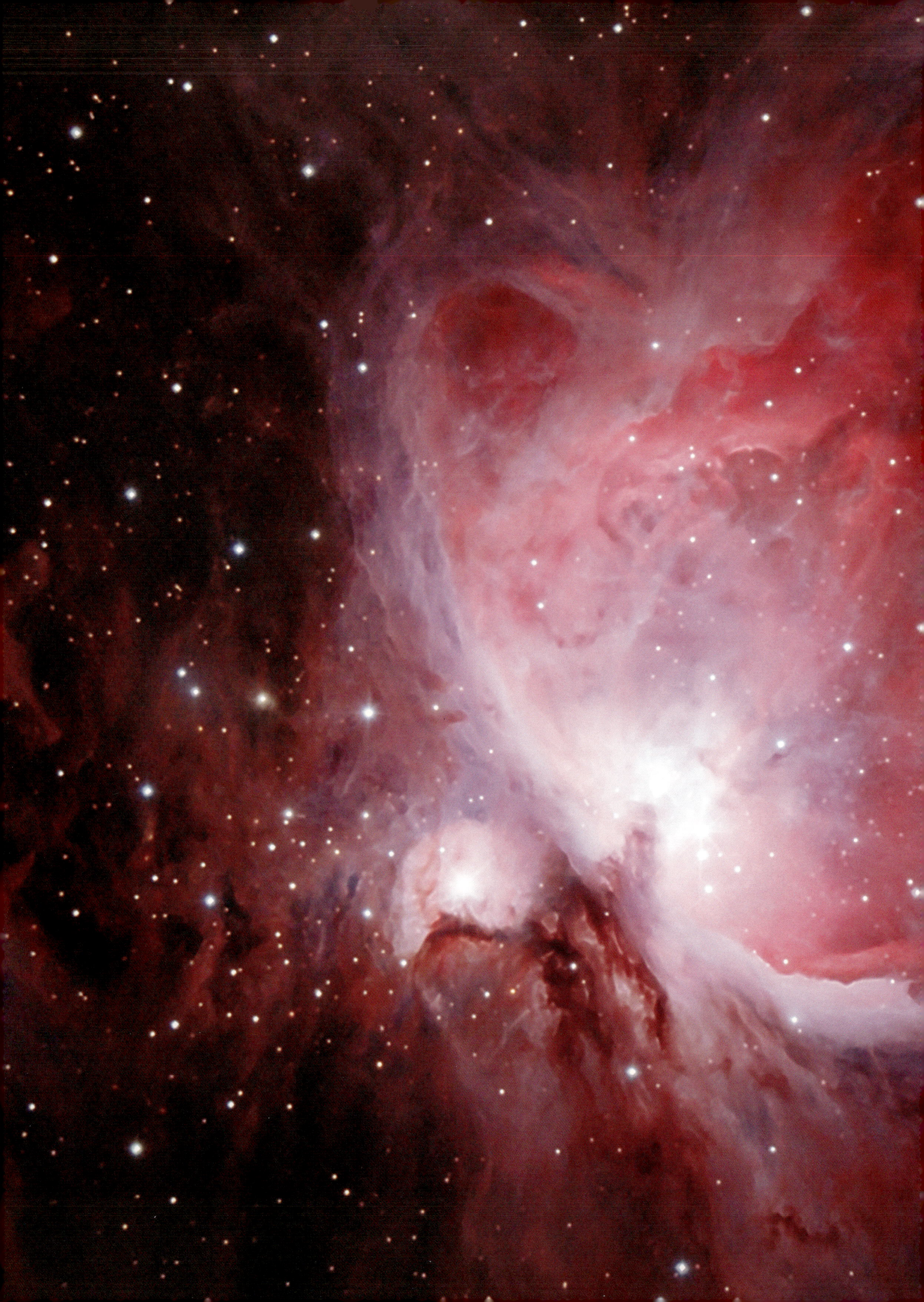

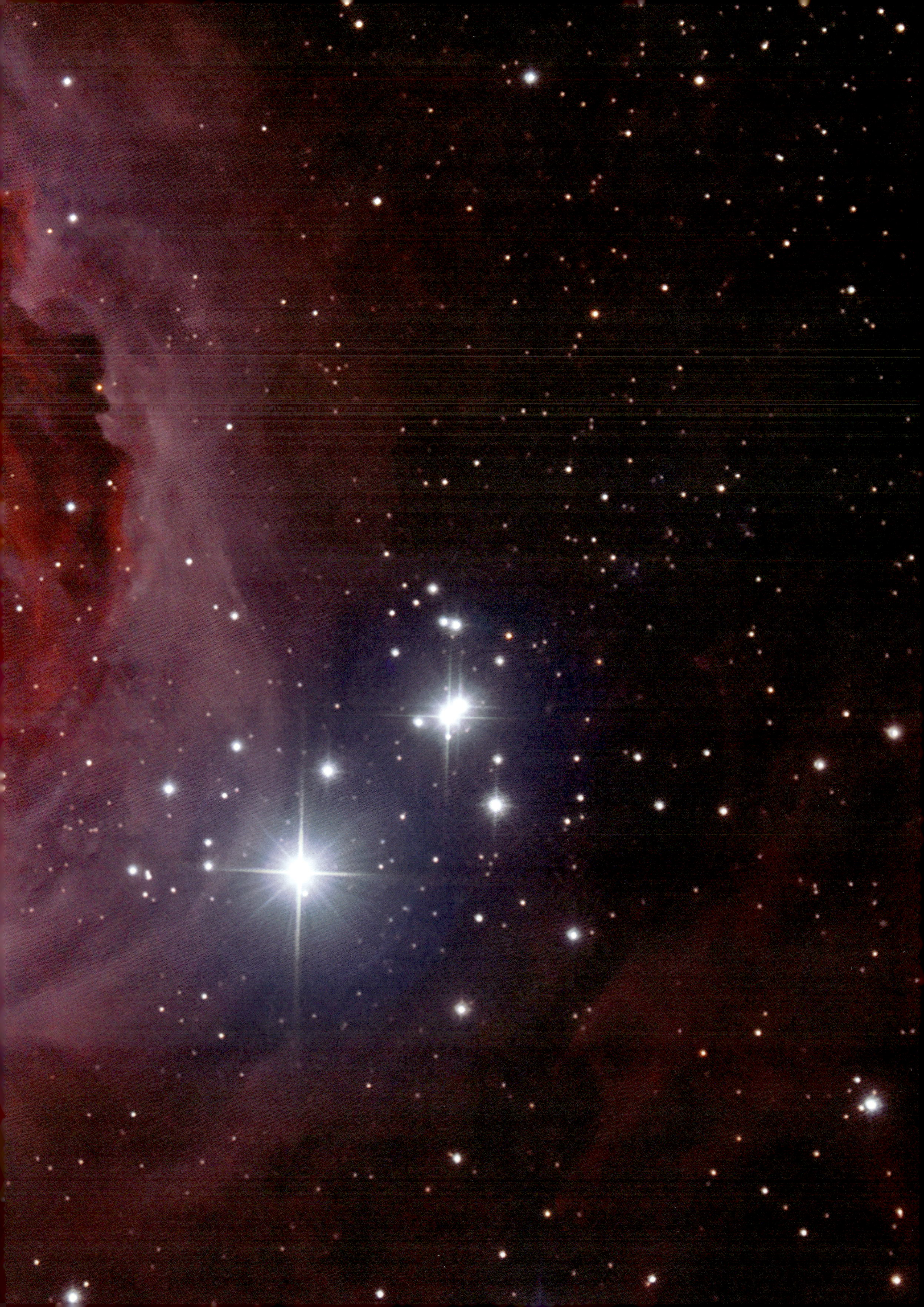

< Previous Two Pages

M43 and M42

This detailed image of the roundish M43 and the larger M42 hints at the three-dimensional shape of the area. Scientists believe that we're actually viewing an opening into a large bubble in space that's been created by the energetic stellar winds of the hot young stars inside.

M43 and M42 are both part of the Orion Nebula, but were originally cataloged separately in an era when only visual observation was done and the two appeared separate to observers in their telescopes.

M43 / M42 image: 2 frame mosaic, 1 hour 29 minutes exposure, SXV-H9C / Hyperstar.

Opposite >

Trapezium Region

At the core of M42 in the very brightest part at high magnification we the brilliant young stars responsible for energizing the nebula and cleari bubble in space. The four brightest stars, lettered A through D, are arrar in a recognizable pattern that gives rise to the name Trapezium. On cl inspection we see that there are two more stars, E and F, that are a dimmer and thus a bit more challenging to see through a small telescope

Trapezium image: 36 minutes exposure, Canon EOS-20D / LX200.

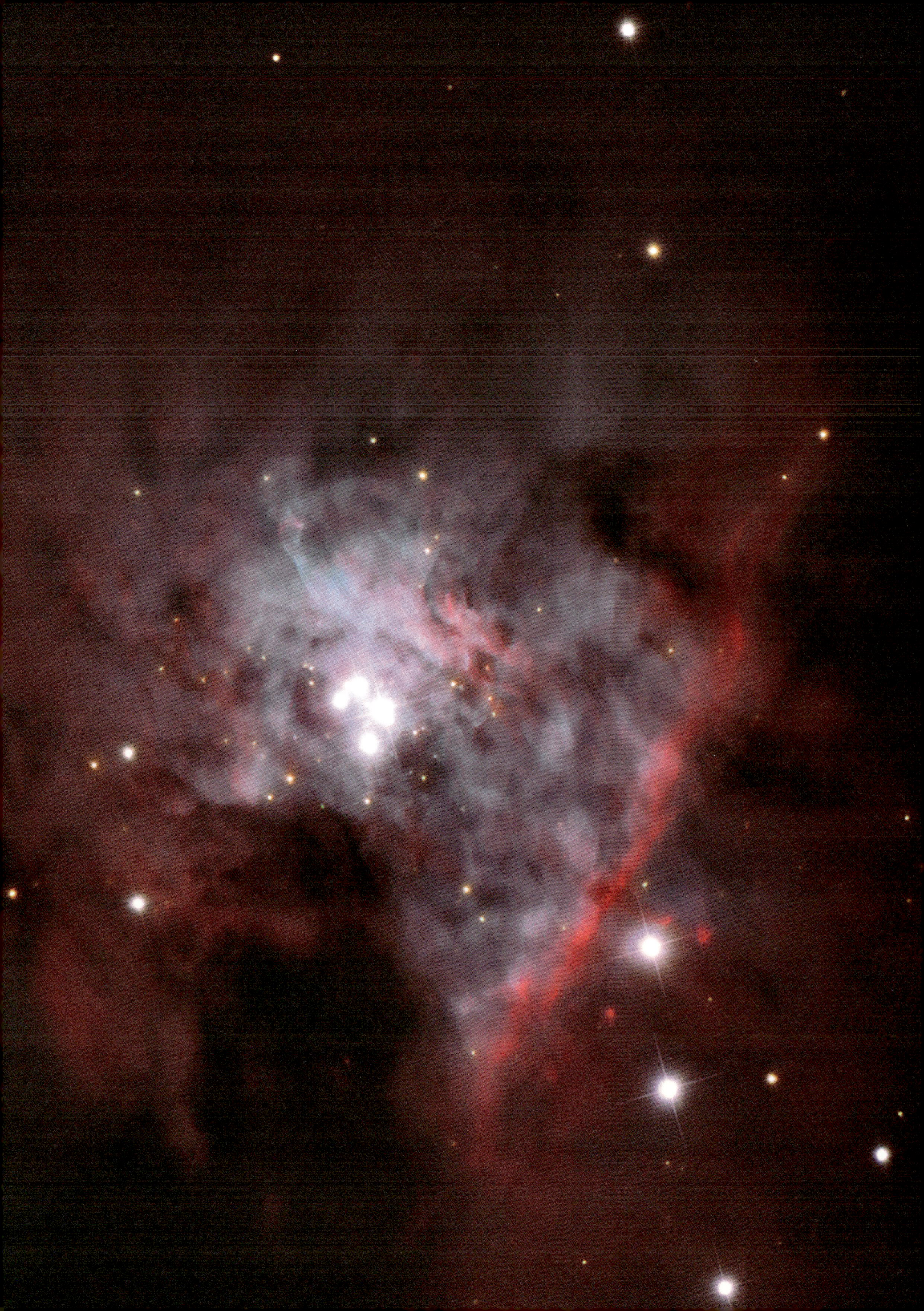

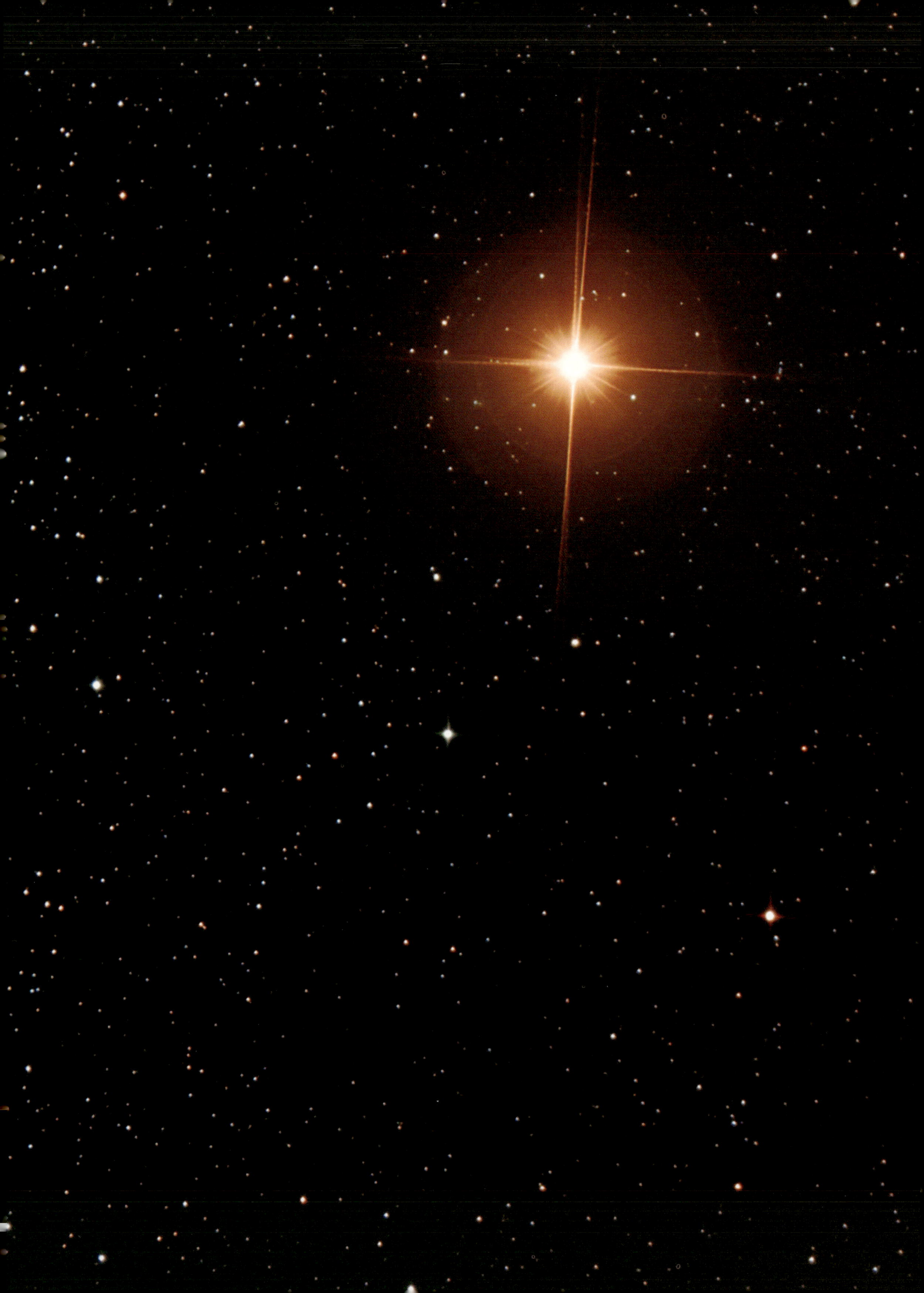

Betelgeuse

At the shoulder of Orion, we find the 9th brightest star in the nighttime sky, the semiregular supergiant star Betelgeuse.

Betelgeuse is one of the largest stars known. It is so huge, at more than 500 times the diameter of our own Sun, that if it were in the position of our Sun its surface would extend past the orbit of Mars, engulfing all the inner planets. Even though its surface temperature is a relatively cool 3,500 Kelvin, it puts out an amazing 60,000 times as much light as our Sun, and at only a few million years of age it has already burned up most of its nuclear fuel. By contrast our own sun has used only half its fuel so far in its 5 billion year life.

Scientists think Betelgeuse is currently fusing helium into carbon and oxygen at its core, and it could go supernova at any time. When it does, at only about 400 light-years distant, it will be visible in the daytime and cast shadows at night. According to one theory, if its rotational axis happens to be pointed directly at us, it could actually bathe Earth in radiation sufficient to diminish our ozone layer. Of course, this is just a theory!

Because of its extreme size and relative closeness to Earth, Betelgeuse is one of the very few stars that have been resolved into more than a pinpoint of light in the most powerful telescopes.

Betelgeuse image: 8 minutes exposure, SXV-H9C / Hyperstar.

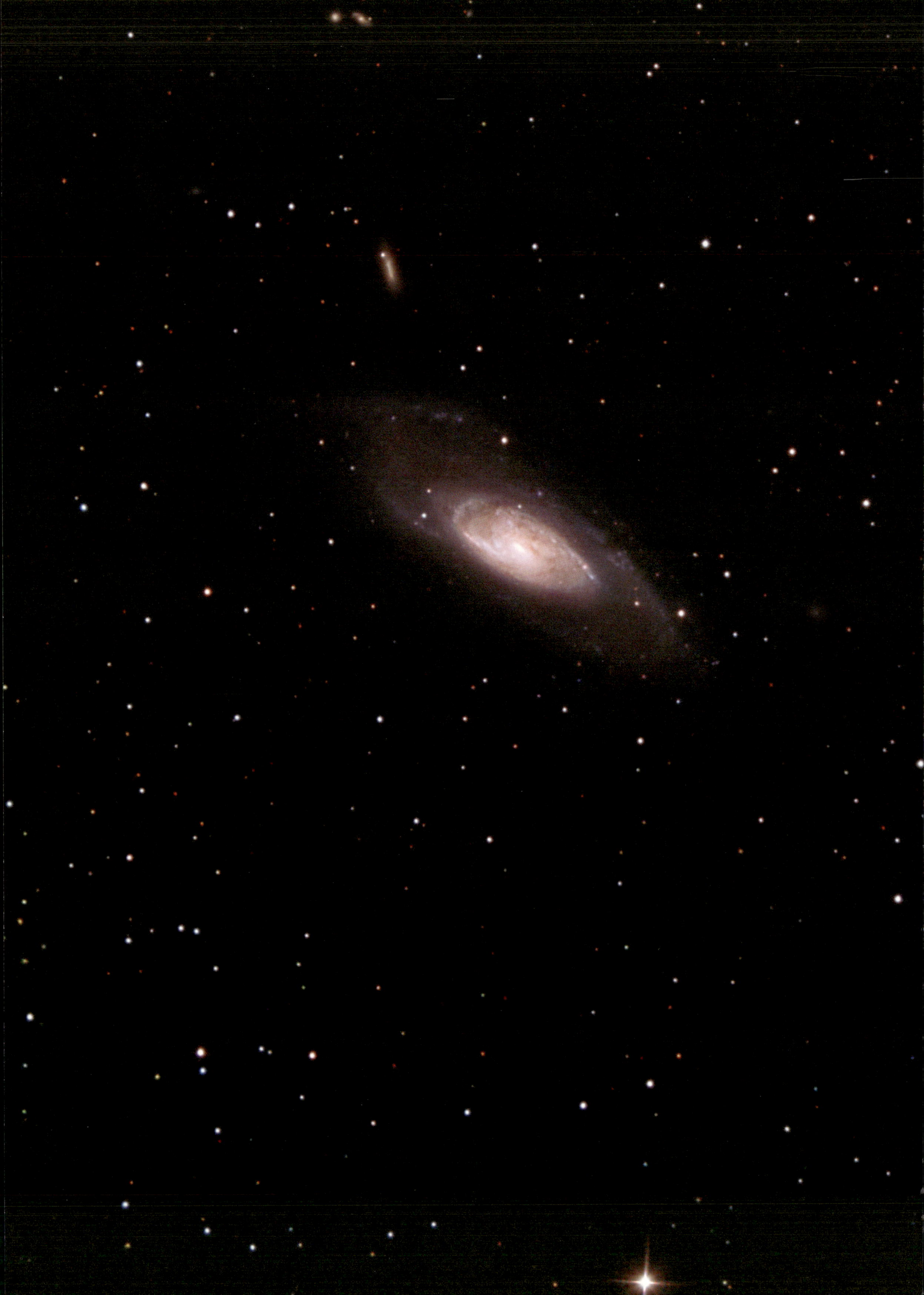

M106 – Spiral Galaxy

Also known as NGC4258 and bright enough to be seen with binoculars in the constellation Canes Venatici, this beautiful and colorful spiral galaxy lies about 24 million light-years from Earth.

At the top of the frame we see galaxies NGC4231 and NGC4232. Above and a bit to the left of M106 we find NGC4248, and to the right we see the small, dim UGC7356. There are several other dim galaxies visible in this image as well.

M106 image: 30 minutes exposure, SXV-H9C / Hyperstar.

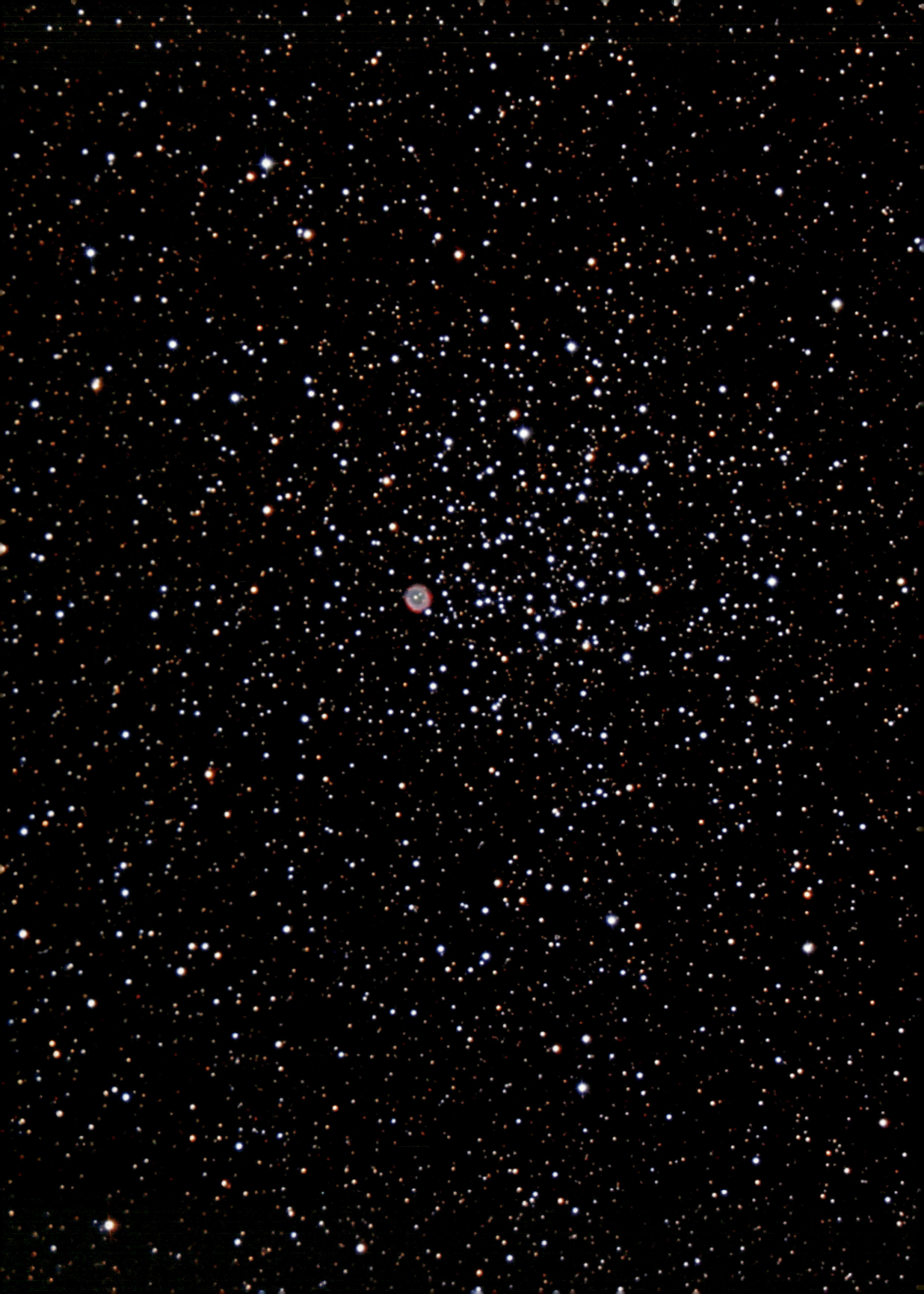

M46 – Open Cluster and Small Planetary Nebula

In the constellation Puppis, we find the beautiful open cluster M46. Its hot, young blue stars dominate the field.

Seemingly within M46 we see a small planetary nebula, designated NGC2438, but precise measurements of its motion reveal that it is moving at a different speed than the stars of the cluster, and it has been estimated to be about 2,000 light-years closer to Earth than the cluster's 5,000 light-year distance. Thus the nebula is not related to the cluster.

This cluster was the first object added to Charles Messier's catalog just after he published his first edition containing objects M1 through M45.

M46 image: 31 minutes exposure, SXV-H9C / Hyperstar.

NGC7023 – The Iris Nebula

About 1,300 light-years distant, in the constellation Cepheus, the Iris Nebula reflects the blue light of a young, hot star. Near the star there is a hint of pink as the material glows reddish in the energetic star's brilliant ultraviolet light.

This image is exposed deeply enough to begin to allow you to see the complex dark masses of interstellar dust that extend out for some distance from the brighter parts of the nebula.

Iris Nebula image: 1 hour 46 minutes exposure, SXV-H9C / Hyperstar.

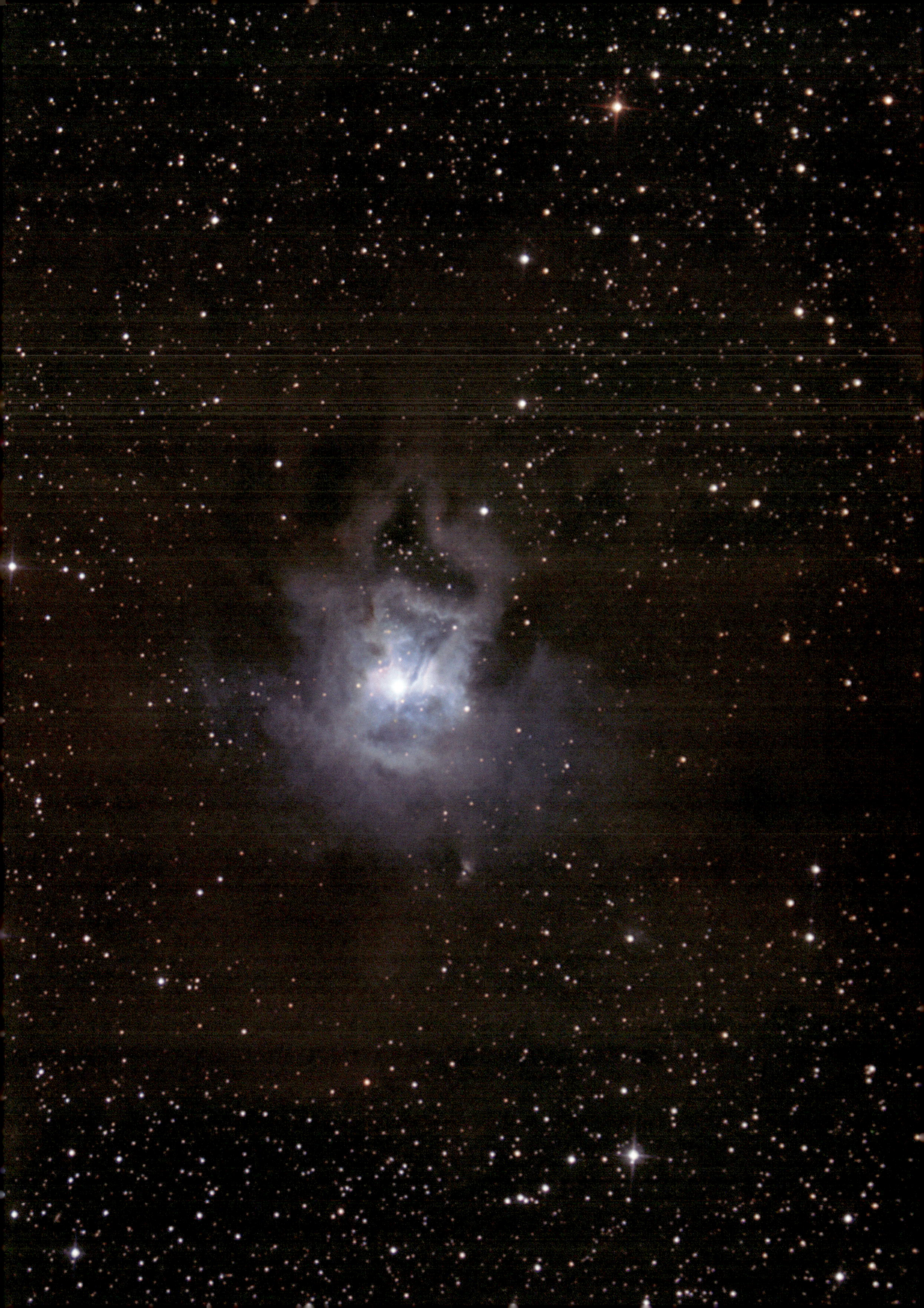

IC5146 – The Cocoon Nebula

An open cluster of stars is just developing in this strikingly beautiful nebula complex in Cygnus.

Note the dark band of gas, called a molecular cloud, which is partially obscuring and coloring the stars from the upper-left part of the image. As a massive new star began to shine, it opened a space in the dark cloud and is energizing the hydrogen gas to glow red as well as lighting up its surroundings with its blue-white light, meaning the Cocoon is at once an emission, reflection, and dark absorption nebula.

Cocoon Nebula image: 2 hours 26 minutes exposure, SXV-H9C / Hyperstar.

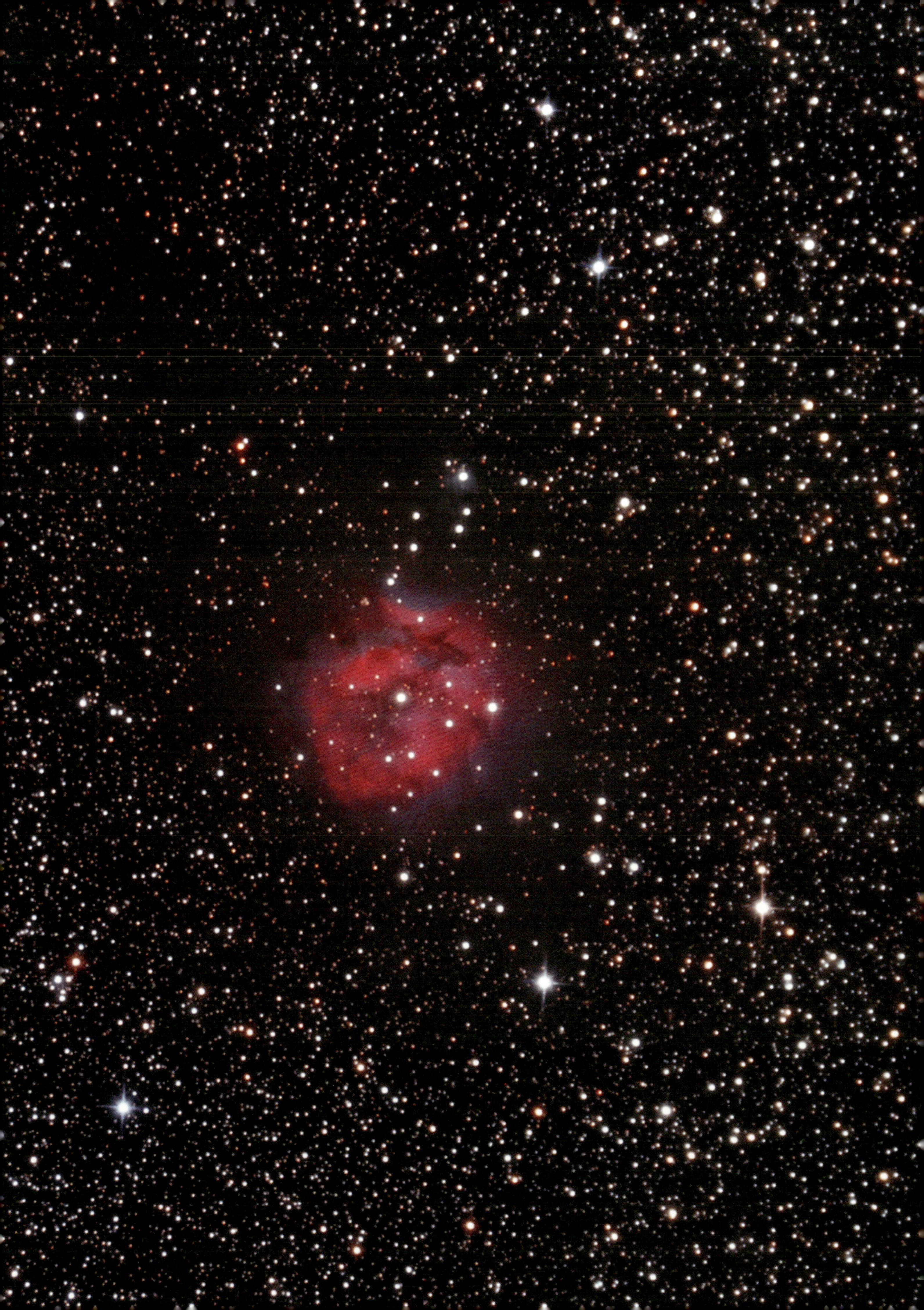

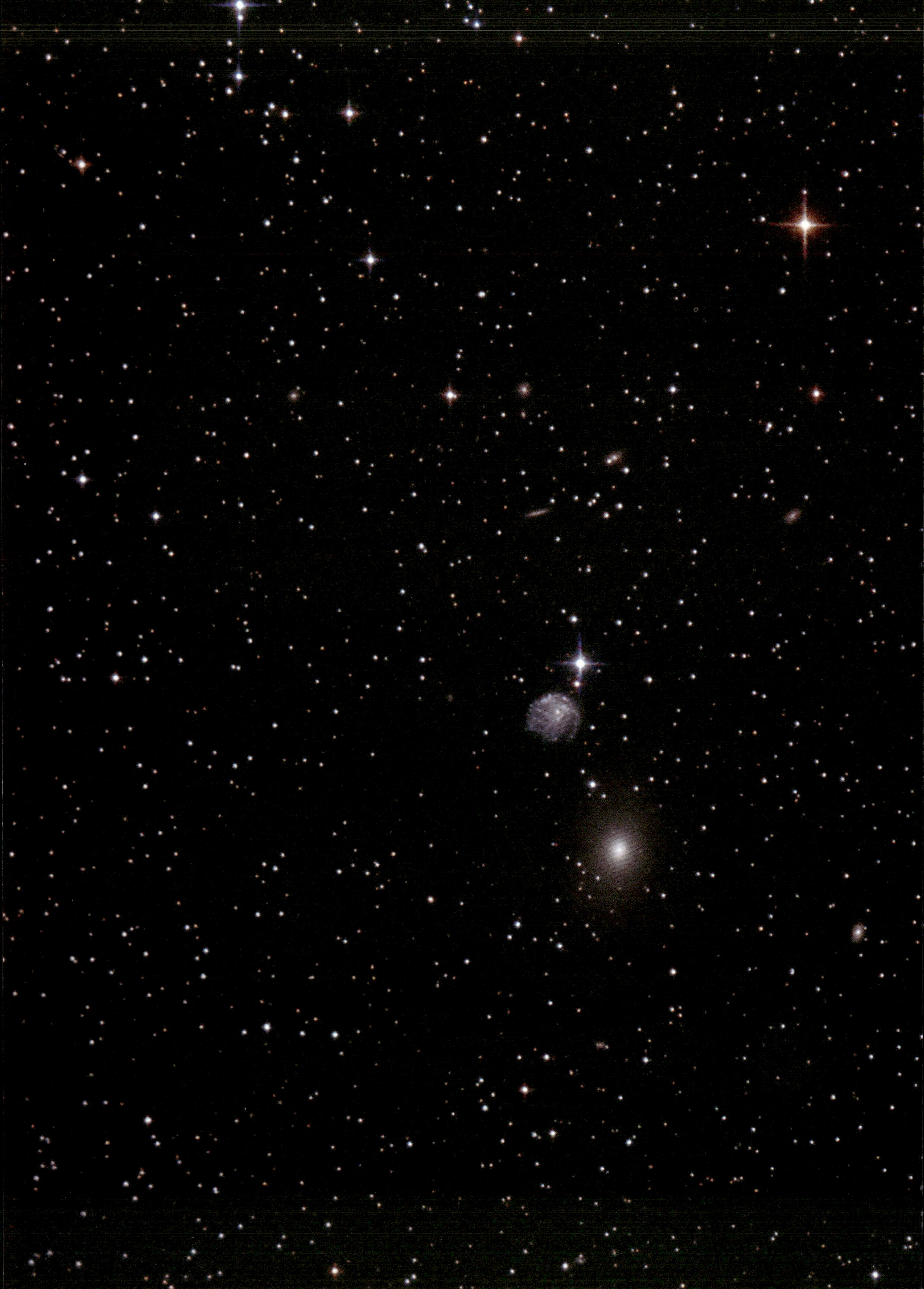

NGC2276 and NGC2300 – The Odd Couple of Galaxies

Most striking in this image is the utterly different appearance of the two largest galaxies. NGC2276 appears as a more bluish concentrated spidery spiral, while NGC2300 has a smooth, lenticular shape and is more yellowish. Though they are very different from one another, scientists believe them to be in the same neighborhood.

What makes this pair particularly interesting is that X-ray observations tell us they are immersed in a massive sea of extremely hot intergalactic gas, which has been credited with causing a high level of star formation in the spiral and which, some scientists say, by its presence serves as observational evidence for the theory of dark matter.

A negative image is presented below to make it easier to see the galaxies.

NGC2276/NGC2300 image: 1 hour 15 minute exposure, SXV-H9C / Hyperstar.

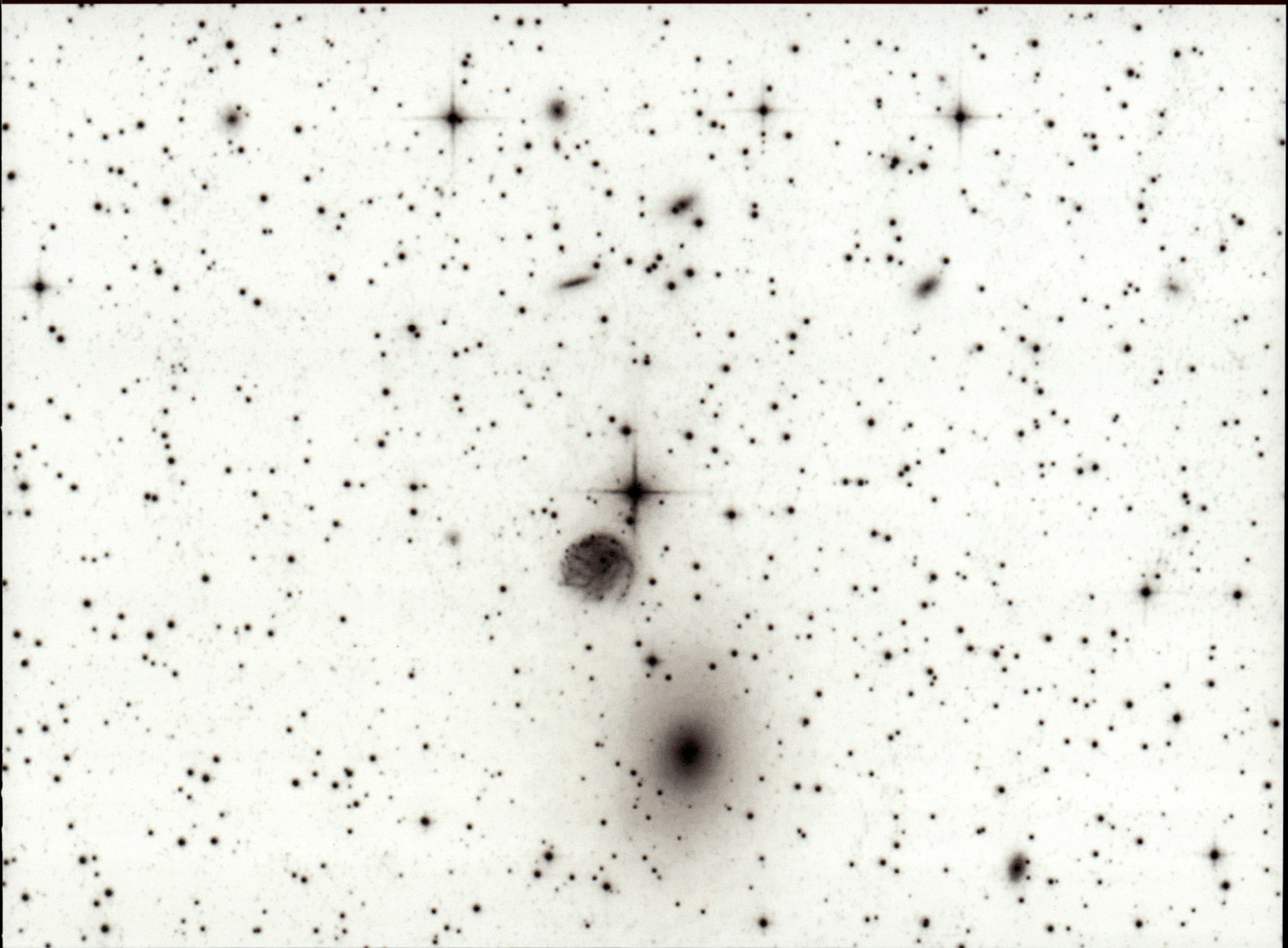

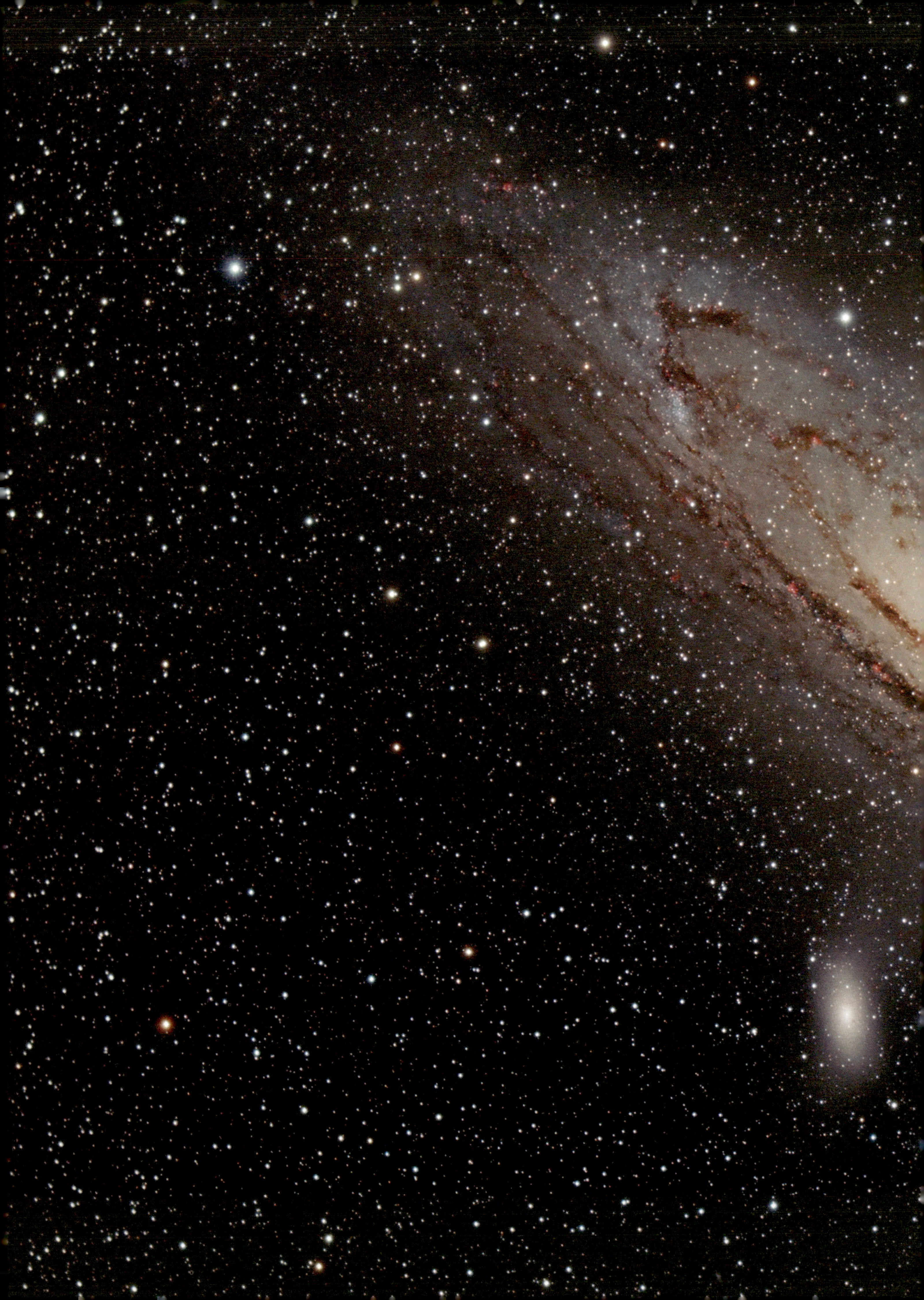

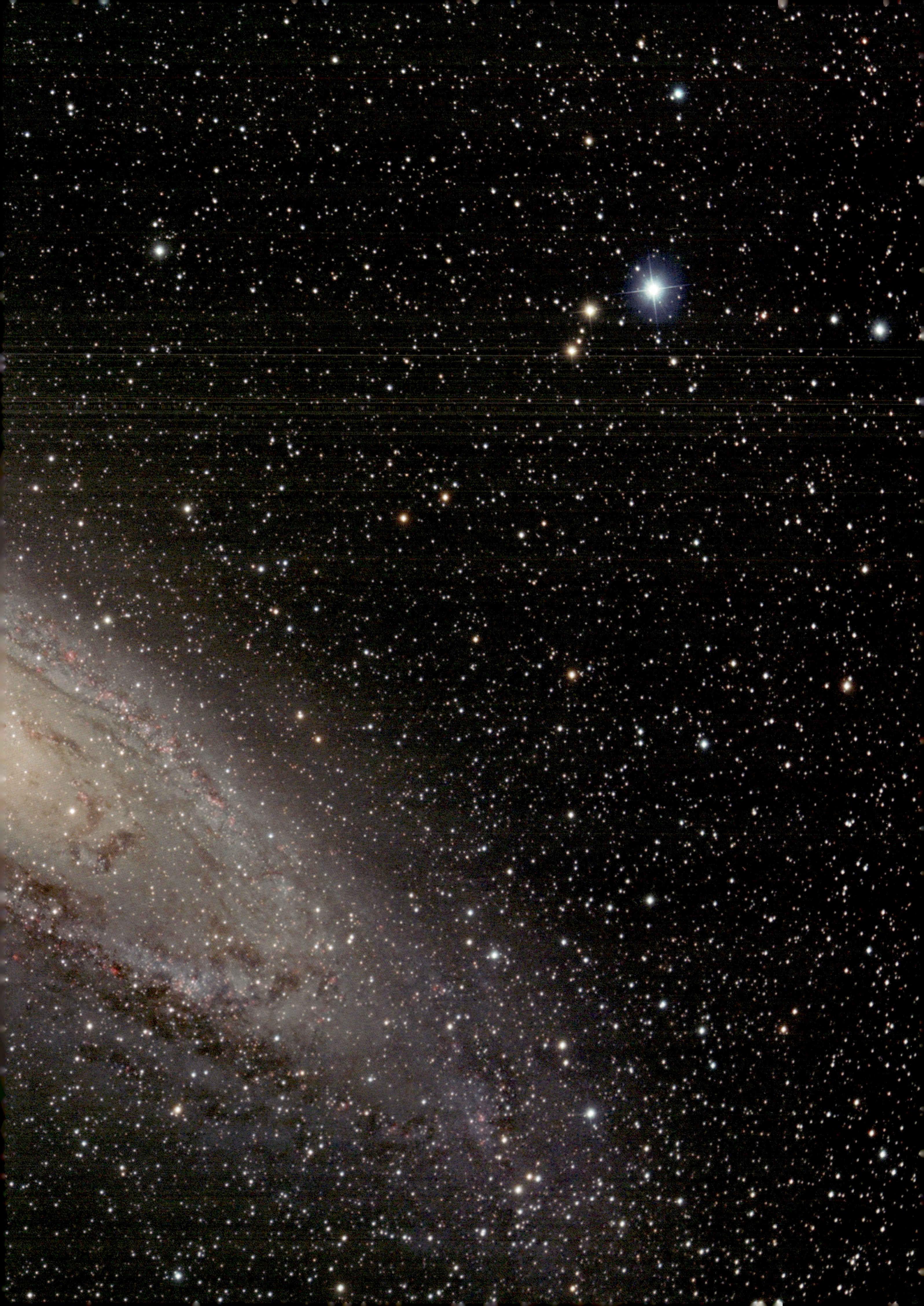

< Previous Two Pages

M31 – The Great Andromeda Galaxy

Seen in the constellation Andromeda, our closest neighbor in the Local Group of galaxies lies at a distance of 2.3 million light-years away and is regarded by many as the most distant object that can be seen (as a faint glowing core) with the naked eye.

The M31 "island universe" including its faint extents is large in the sky, covering over 3 degrees - nearly seven times the diameter of a full Moon! With the soft combined light from its estimated one trillion stars, you can see prominent dust lanes in the spiral structure, Ha and HII emission nebula regions, and even some star clusters.

In this image we can also see M31's companion galaxies M32 (above) and M110 (below).

This Sky 90/M25C single frame image places M31 along the camera diagonal to just fit the whole galaxy in the frame and is comprised of both RGB color and Ha red filtered exposures. The hundreds of Sky 90 exposures that went into this image were collected in many imaging sessions over two years, and there is also some earlier Hyperstar data of the core region included as well. Total exposure time for these images exceeds 30 hours!

Opposite >

A Clearer View

What would M31 look like without the intervening bright stars of our galaxy in the way? If you were able to travel just outside our galaxy intergalactic space and then view the Great Andromeda Galaxy, it would something like the lower image here. Note that the small "star-like" l remaining are actually cataloged globular clusters in our closest gal neighbor!

The gravitational connections of M110 and M32 with M31 are easier to here. Note the slight haze from the millions of stars that have been p between the galaxies by gravitational forces.

Stars digitally removed, known clusters retained.

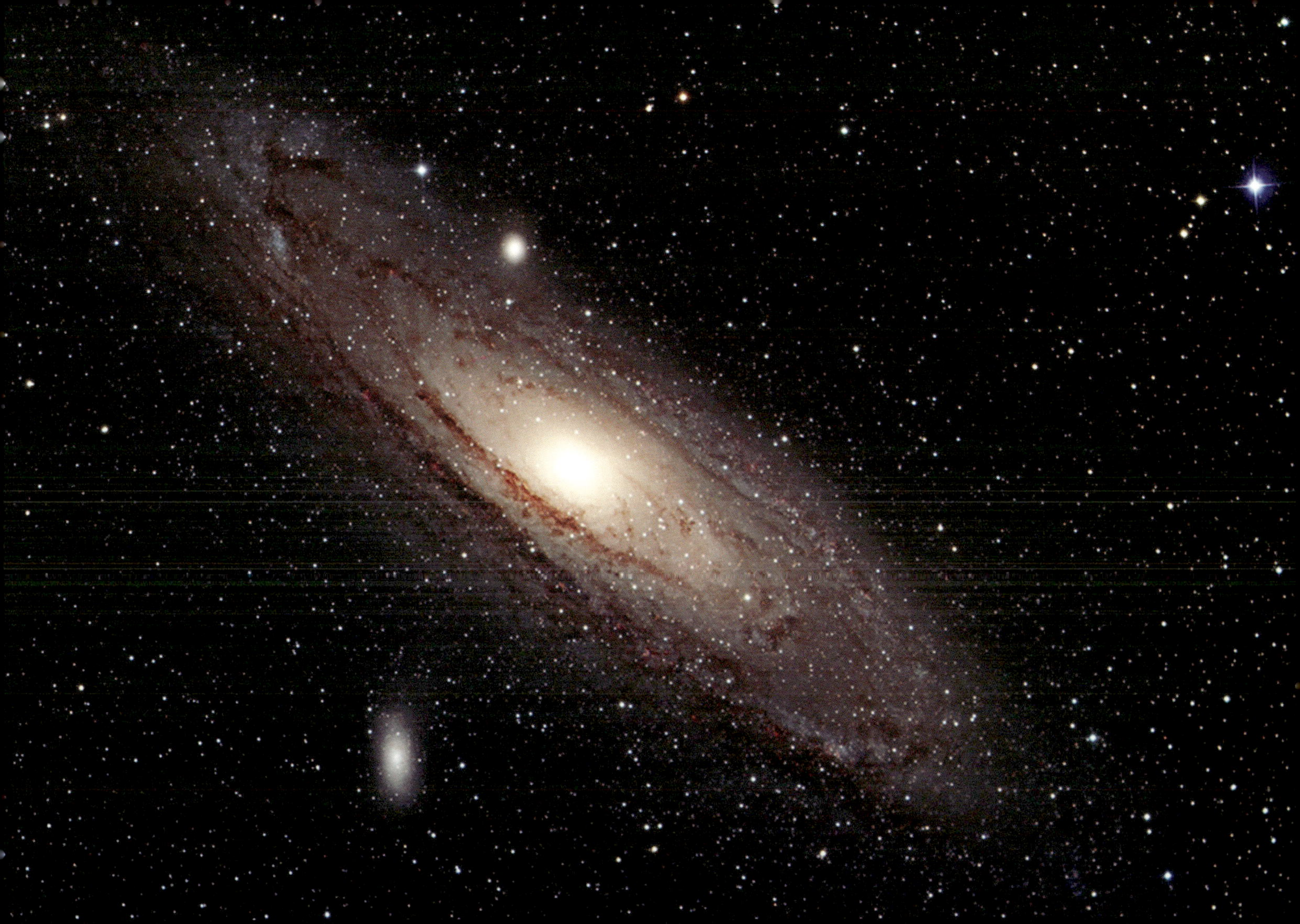
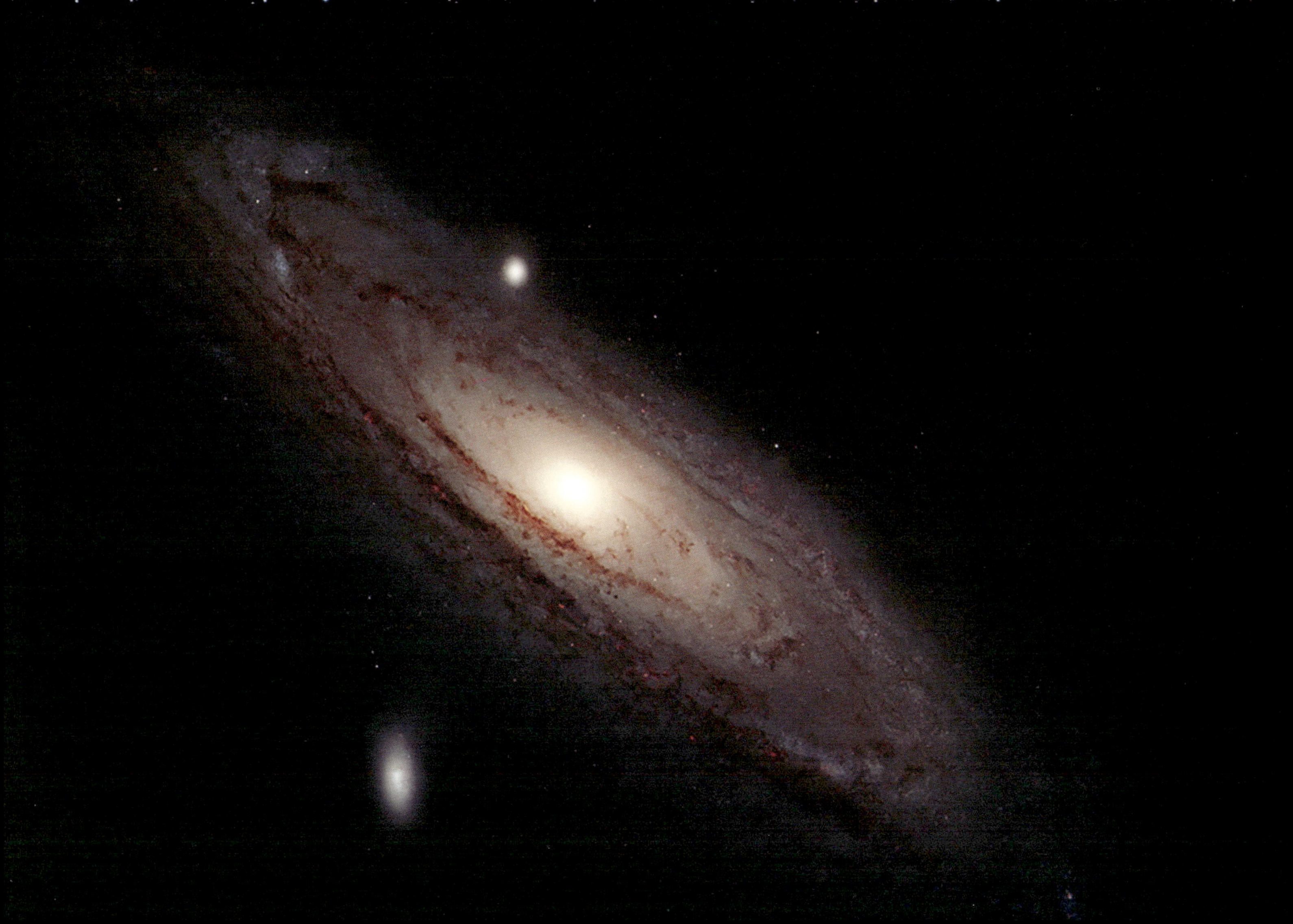

M44 – Praesepe, the Beehive Cluster

This beautifully colored open cluster of stars, Messier 44, also commonly known as Praesepe and the Beehive Cluster, lies in the constellation Cancer. This cluster can be seen with the naked-eye in reasonably dark locations, appearing as a faint glowing patch of light.

Known since antiquity, M44 (also cataloged as NGC2632) lies at a distance of 515 light-years and appears 1.2 degrees in diameter overall with a brightness of magnitude 3.1. It is a strange fact that the stars that make up Praesepe ("the Manger") are actually brighter than the stars that make up the constellation in which it resides!

Beehive Cluster wide-field image: 2 hours 53 minutes exposure, SXVF-M25C / Sky-90.

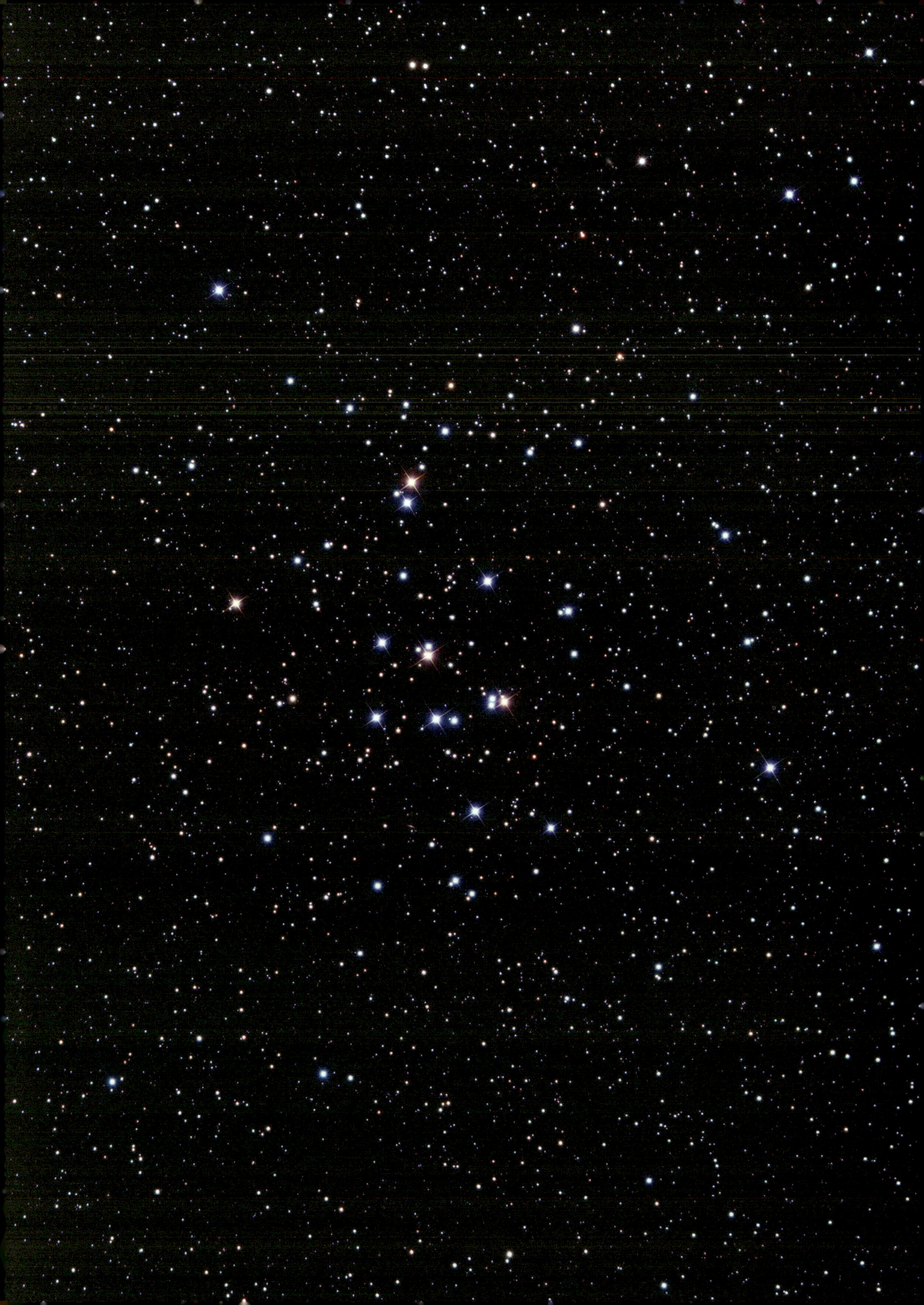

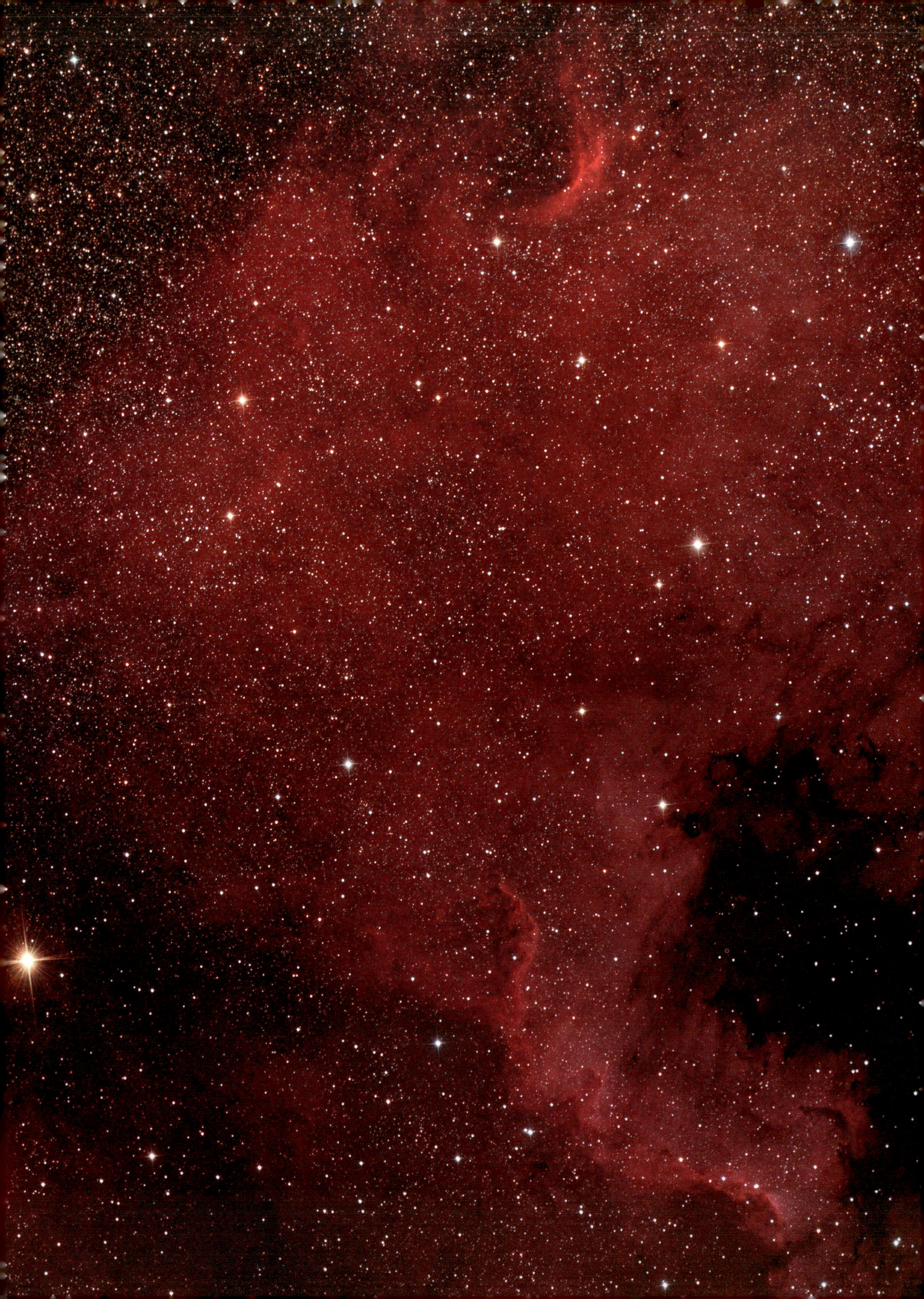

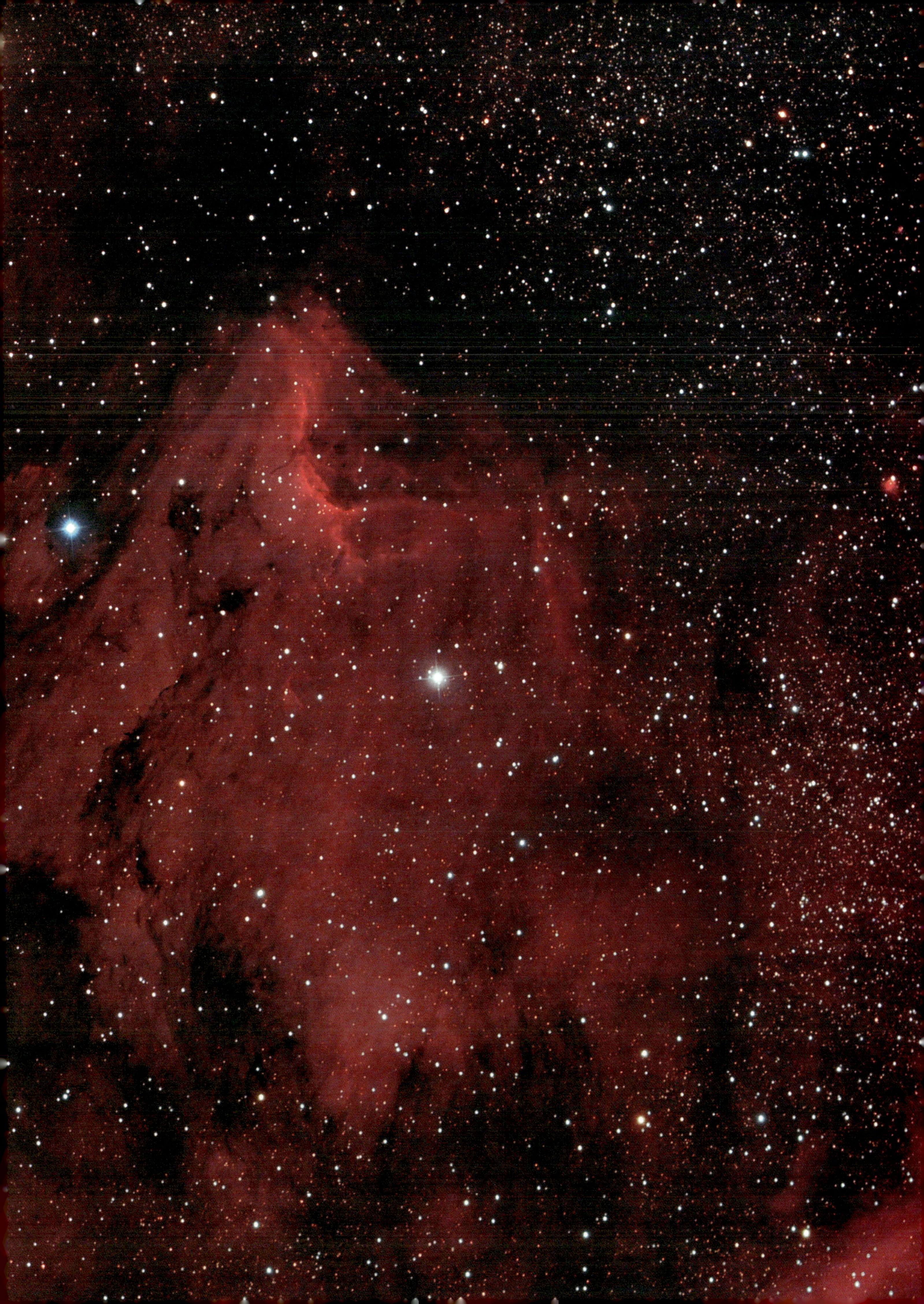

< Previous Two Pages

NGC7000 and IC5070 – North America and Pelican Nebulae

In these images, we see a very large and rich area of emission nebulosity in Cygnus near Deneb at the tip of the cross. It's no wonder these nebulae have been given such descriptive names. Some have even labeled the small nebula between the two giants near the "nose" of the Pelican as the Skull Nebula.

It's not known which star is causing the formation of so much glowing gas, but most astronomers agree that it's an intervening band of interstellar dust that's blocking light and giving the nebulae their interesting shapes.

North America Pelican image: 9 hours RGB exposure, SXV-M25C / Sky-90 and Hyperstar III.

Opposite >

Cygnus Wall Close Up

What would be the Central America area in the North America Nebula sh on the previous two pages, the Cygnus Wall is a spectacular front of de gas and dust whose surface is both illuminated and eroded by hot stars.

Cygnus Wall image: Two frame mosaic, 2 hours RGB exposure, SXV-H9 Hyperstar.

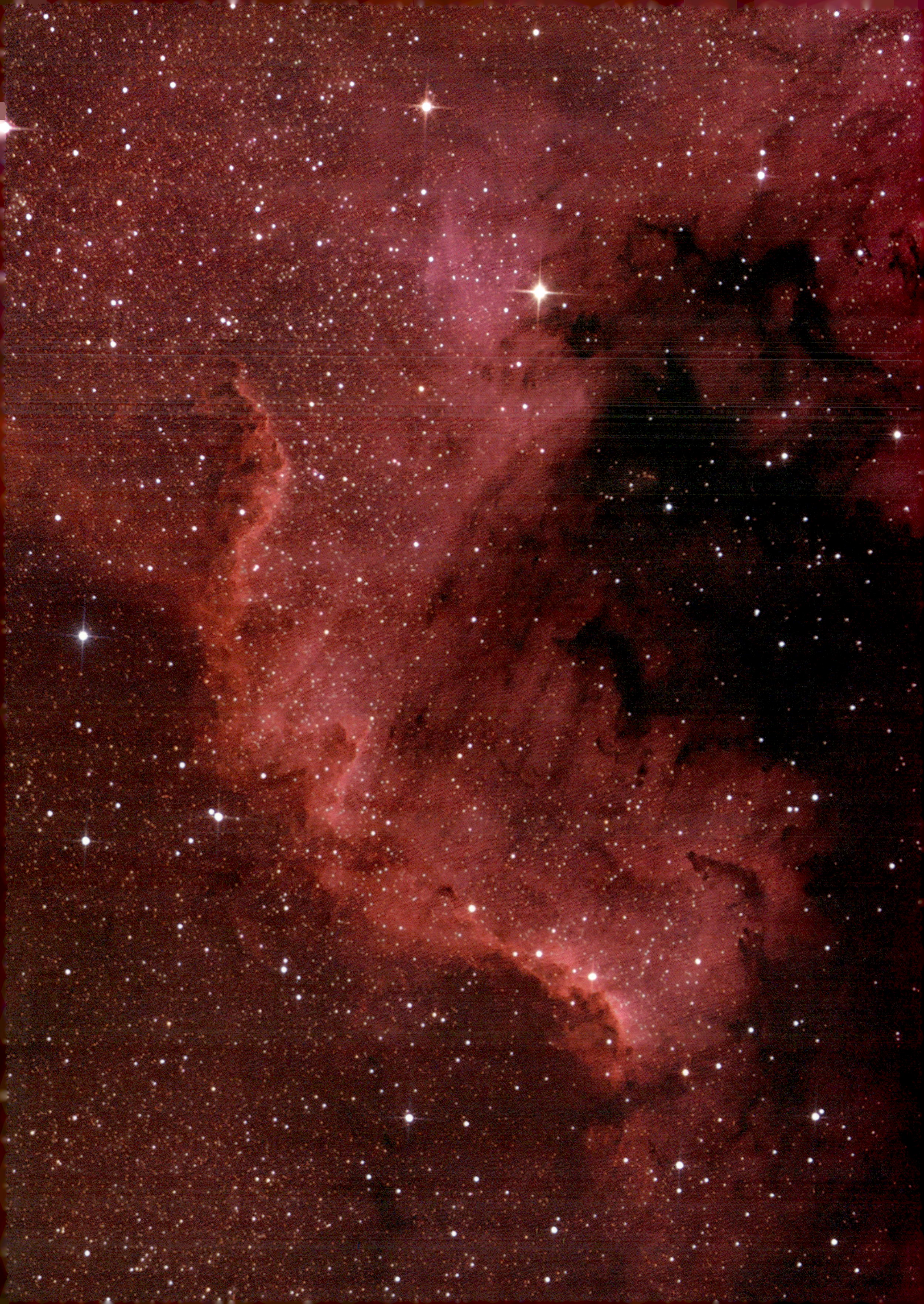

Abell 2065 – The Corona Borealis Galaxy Cluster

At the time this galaxy-cluster was imaged, this was the most distant group of objects we had knowingly photographed using the SXV-H9C camera. The galaxy group Abell 2065 lies at a staggering 1.5 billion light-years distance in the constellation Corona Borealis, the Northern Crown.

Most of the small objects without diffraction spikes near the top of this image of the Abell 2065 group are galaxies, not stars. The brightest galaxies in this image are magnitude 15.5, while the rest are magnitude 16 and fainter. The whole galaxy group is only about half a degree in diameter overall, so we needed the higher magnification imaging system to provide a reasonable field of view. Recall that the field of view of the SXV-H9C using the Hyperstar lens is about one degree by three-quarters of a degree.

Abell 2065 image: 1 hour 10 minutes RGB exposure, SXV-H9C / Hyperstar.

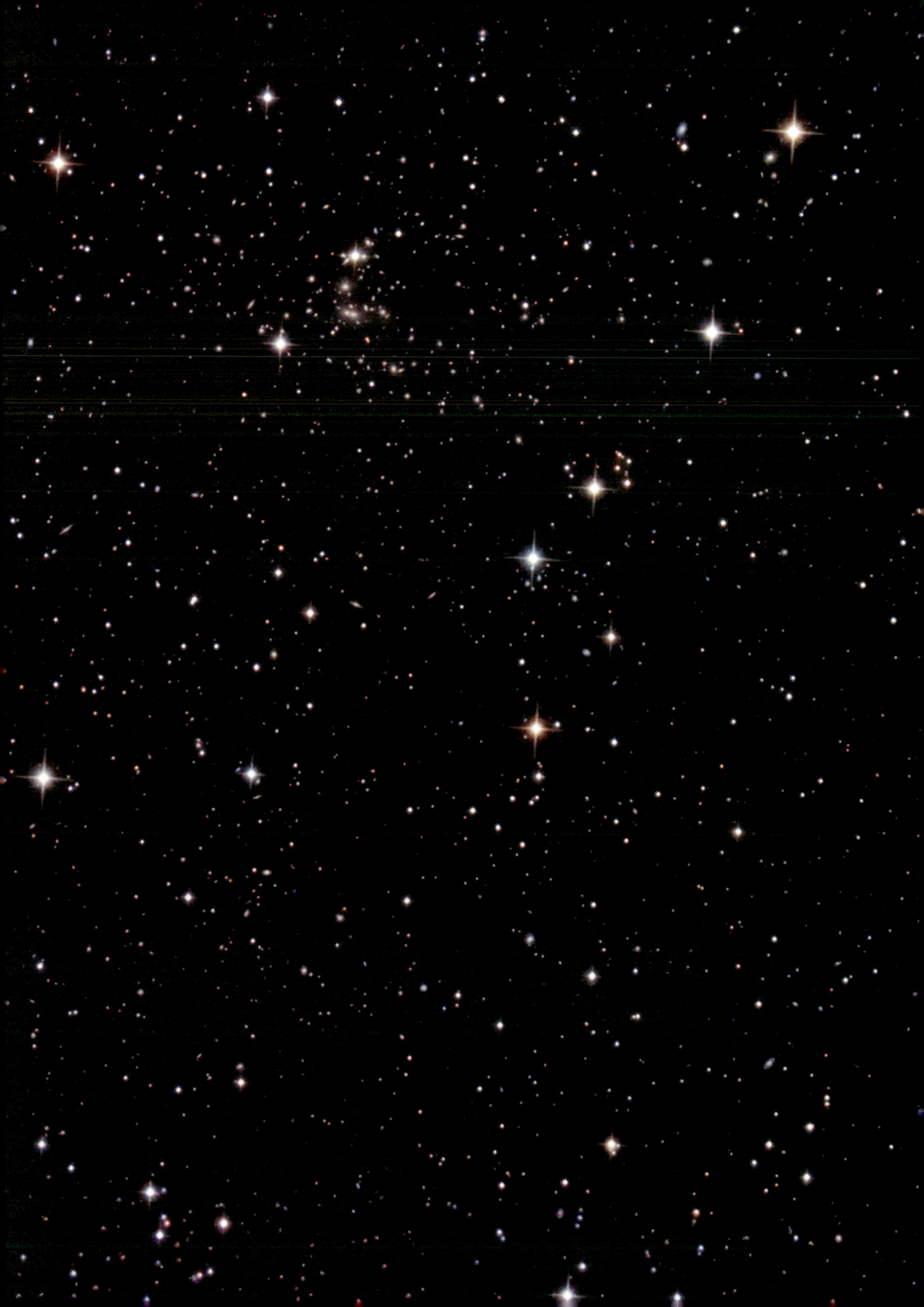

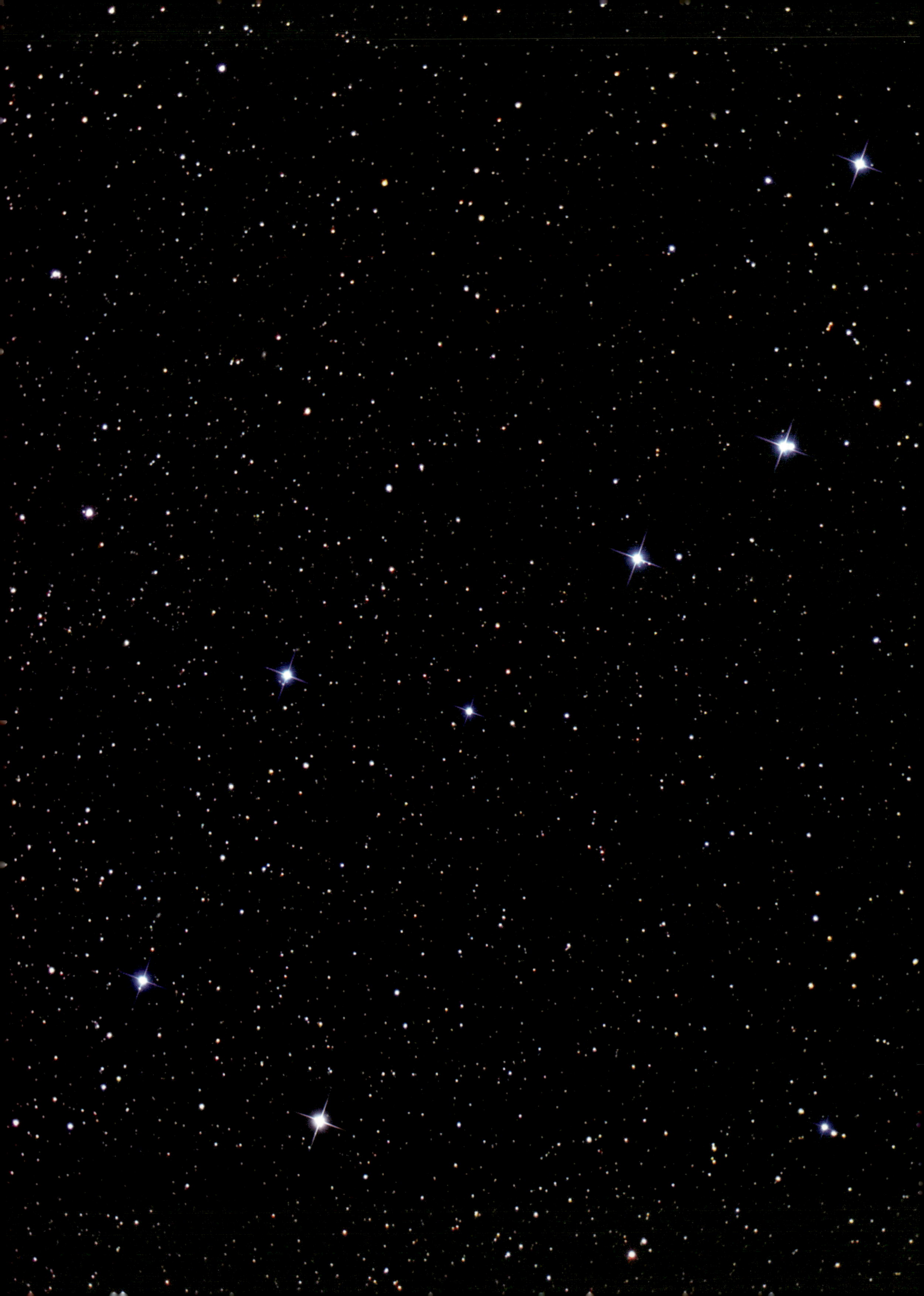

The Big Dipper or Plough

Northern hemisphere observers will recognize this asterism, part of the Ursa Major constellation, which is visible in the northern sky for a large part of the year (depending how far north you are). The two stars at the end of the pan (lower left in this image) currently point to Polaris (not shown in this image). Although North Americans usually call it the Big Dipper, residents of the U.K. more typically refer to this asterism as the Plough.

If you look closely you can see that one of the handle "stars" is actually a double: Mizar and Alcor.

What's not commonly known is that six of the dipper stars (excluding the top-most and bottom-most in this image) appear to be part of a nearby open star cluster, commonly thought to be moving in the same direction and speed as part of the Ursa Major Moving Group.

Big Dipper image: 7 minutes exposure, Canon EOS-20D / 17-40 zoom lens.

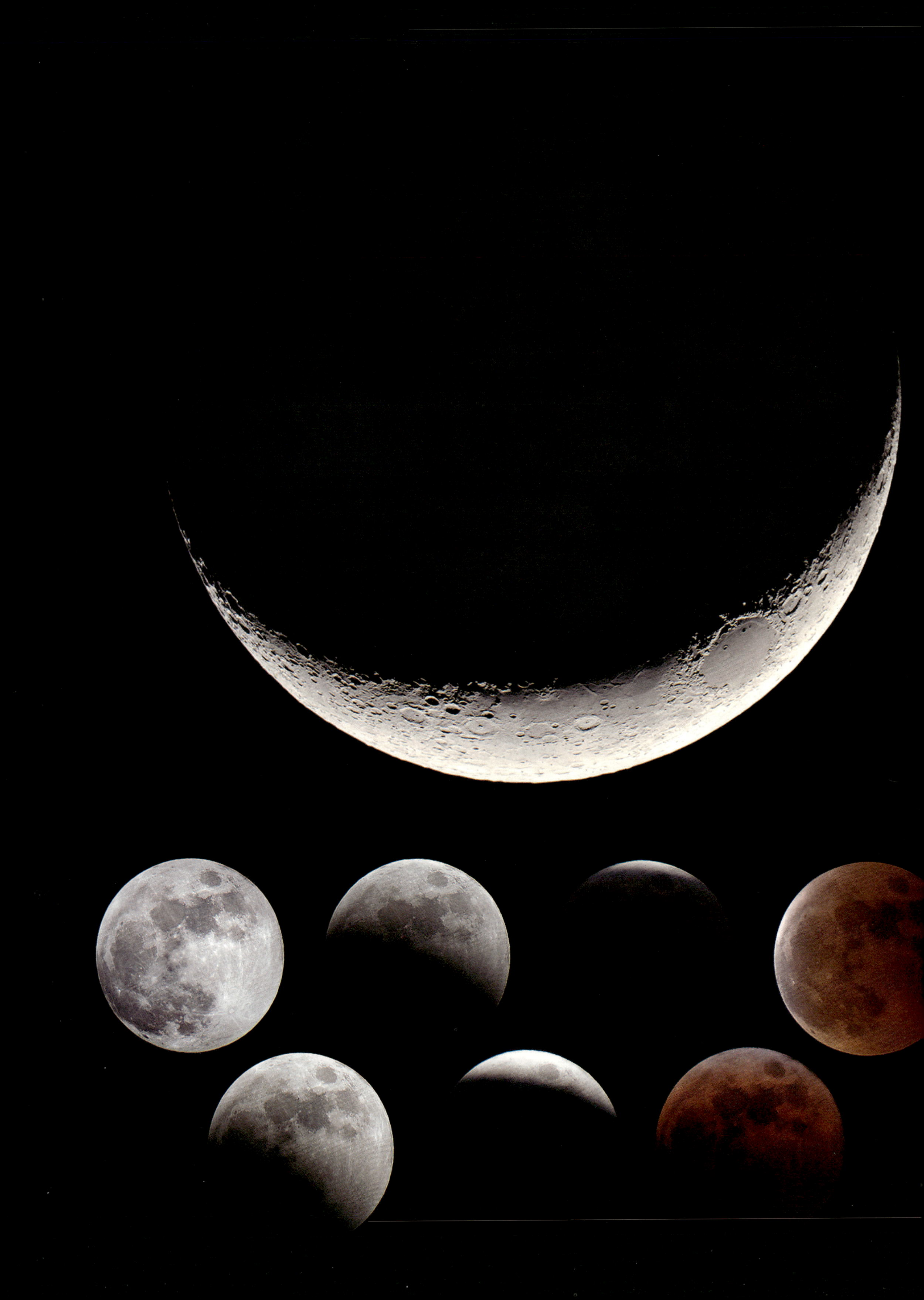

pposite

he Moon - Many Faces or One?

Moon orbits Earth, and although it always shows us the same side, we it in many different lights, depending on its position relative to the Sun Earth.

n the Sun is mostly behind the Moon, we see the Moon as a thin gray-e crescent, the darker part lit up by faint Earthshine - sunlight that has nced off Earth, then the Moon, then back to our eyes. However, when the n is nearly opposite Earth from the Sun it becomes fully illuminated. asionally it lines up just right, and we see a lunar eclipse. At the darkest es of the eclipse the lunar disk is lit only by sunlight that has traveled nd Earth, passing through our atmosphere and lending the Moon a ish-orange glow much like that of a sunset.

larger image on top is of a typical crescent Moon, a few days after the Moon phase, and was assembled from long and short exposures to tate capture of the huge range of brightness levels from the sunlit area ne part lit by Earthshine.

small sequence of images below is of a lunar eclipse that occurred on ober 27, 2004. The bright, full Moon exposure time was just $1/160^{th}$ of a nd, while the fully eclipsed Moon required the EOS-20D shutter to be n for 1 full second.

Next Two Pages >

Is the Moon Really Just Gray?

Other than when in eclipse or near the horizon, the Moon looks pretty gray. Its surface averages about the color and reflectivity of an asphalt highway.

But is it really colorless?

With highly accurate exposure data, we can digitally enhance an image of the Moon so as to more clearly visualize the various materials on the surface. This allows us to see that the surface is composed of many different kinds of rock, including dark, metal-rich lava flows from billions of years ago, light ejecta from huge impacts, and a gradually accumulating covering of fine meteoritic dust, which has muted the colors and given the Moon its overall grayish color.

This high-resolution Moon image was stitched from 15 separate 1/5 second EOS-20D exposures through an LX200 telescope with a focal length doubler. Contrast and color saturation were boosted a large amount digitally in the second image to dramatically highlight the varying composition of the lunar surface.

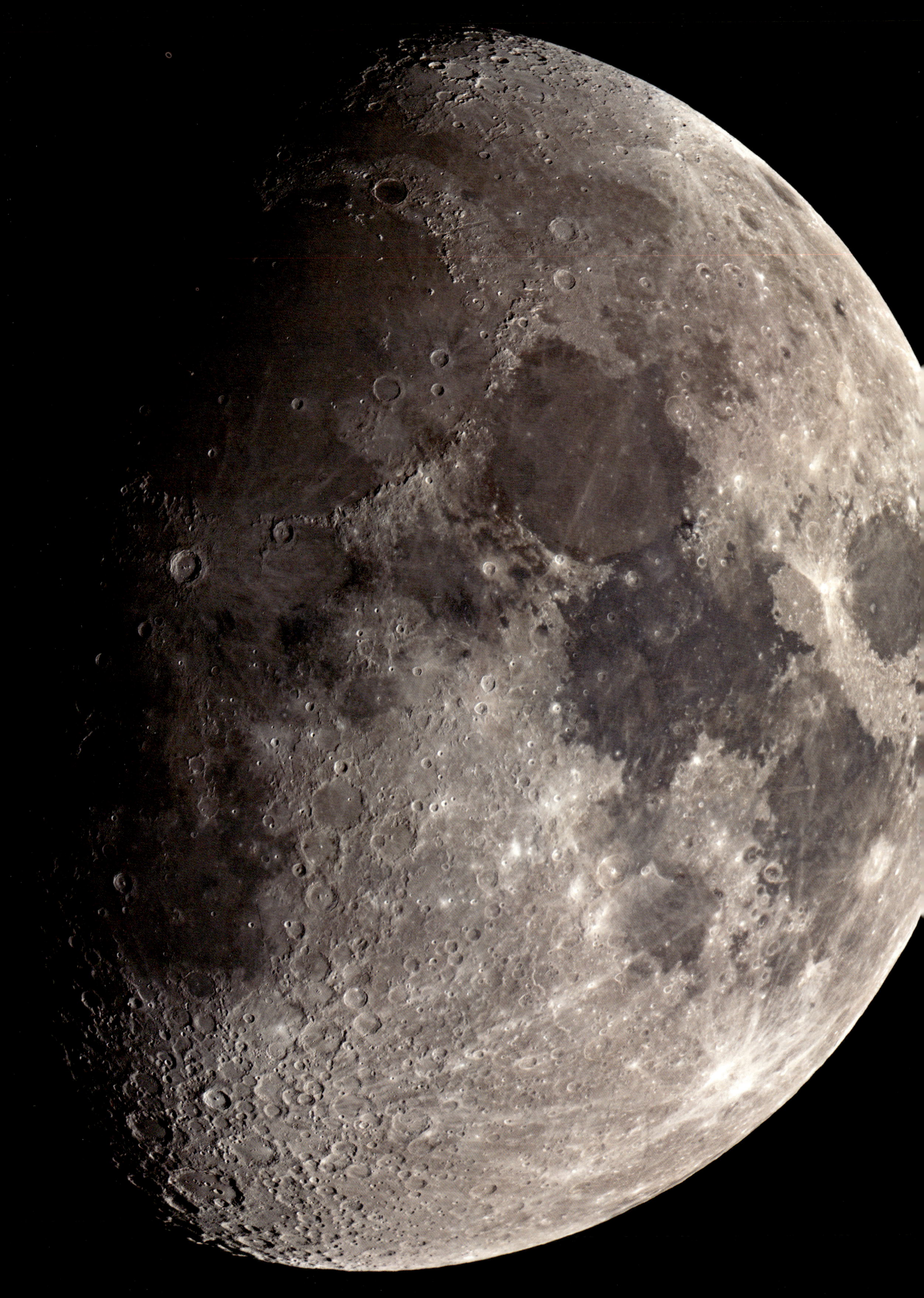

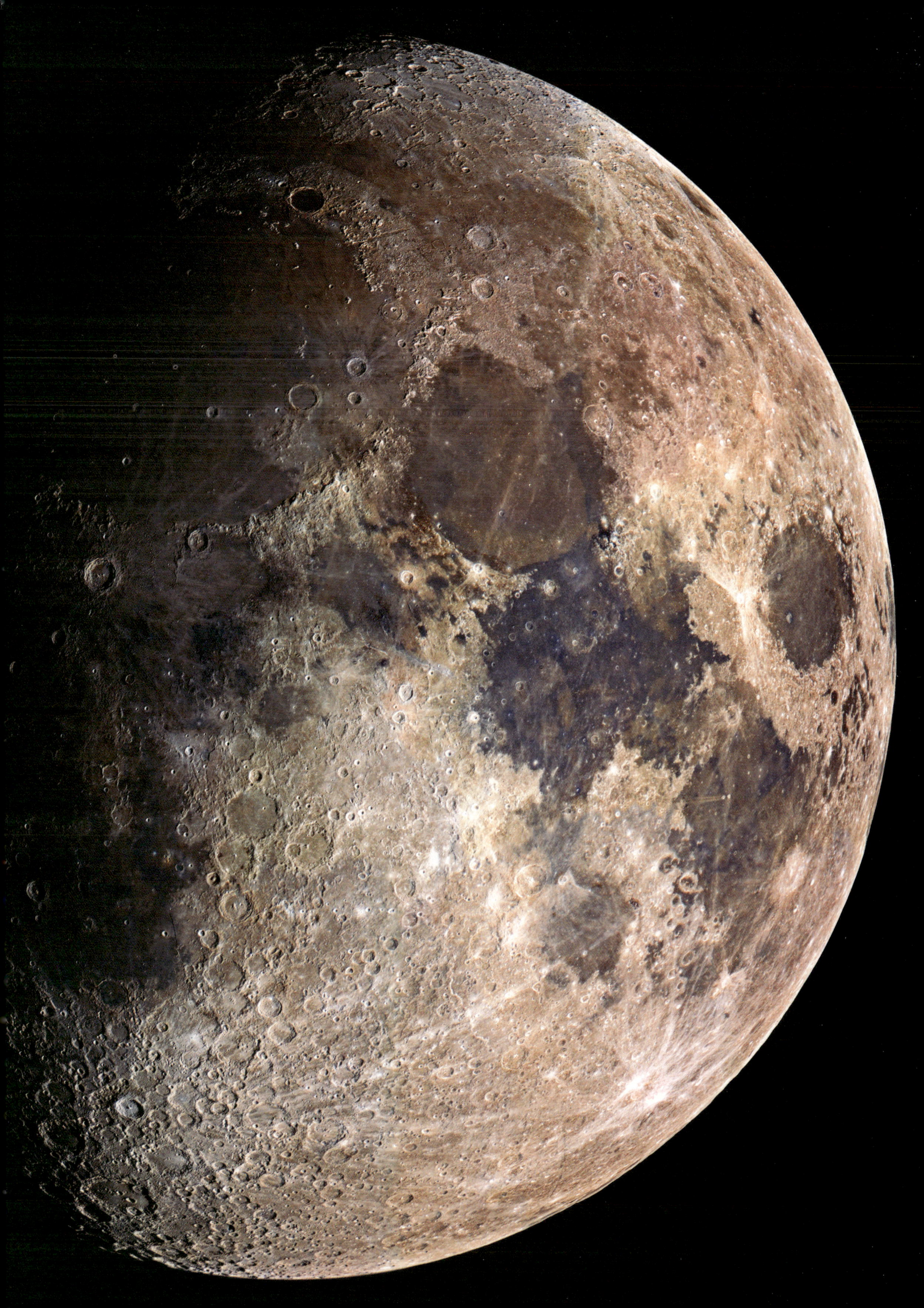

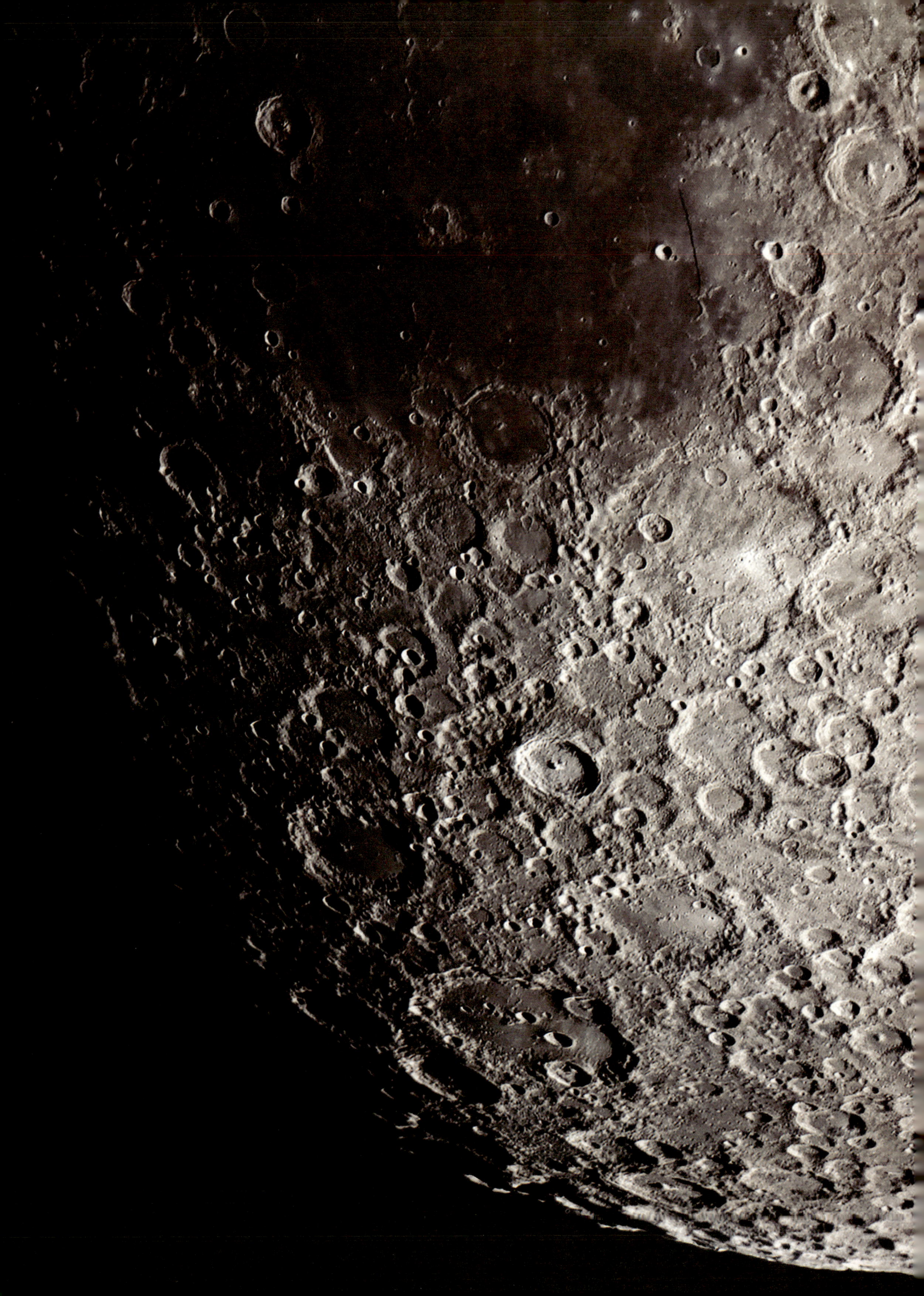

The Moon's Southern Polar Region

With greater magnification we can begin to appreciate the extreme meteoritic bombardment the Moon must have suffered in its early years. Note that Mare Nubium – the smoother portion toward the top-left of this image – has far fewer craters than its surroundings. You can see some that appear to have been filled in. Between 1 and 4 billion years ago volcanic eruptions spewed molten rock over the surface to form the lunar maria, smoothing out the oldest craters and leaving the surface dark, with rocks of iron-rich composition. Scientists believe the Moon is now completely solidified.

Of particular note is the crater Clavius, near the bottom-center of the image. Note the crescent line of craters of decreasing size.

Also, the dark straight line at the upper center-right is known as Rupes Recta, or the Straight Wall. Though this line seems to indicate an abrupt crack, it's actually a gentle drop of about 1,000 feet over nearly a mile in width.

South Polar Region image: Mosaic of 1/5 second exposures, EOS-20D / LX200.

The Northern Moon

Dark crater Plato and the long, thin Alpine Valley highlight this image of the northern regions of the Moon. At bottom-left is the crater Copernicus, which was formed by an extremely violent impact about 1 billion years ago.

Just below and slightly to the right of center is Hadley's Rille (very tiny in this image), which was the landing site of the *Apollo 15* mission.

North Polar Region image: Mosaic of 1/5 second exposures, EOS-20D / LX200.

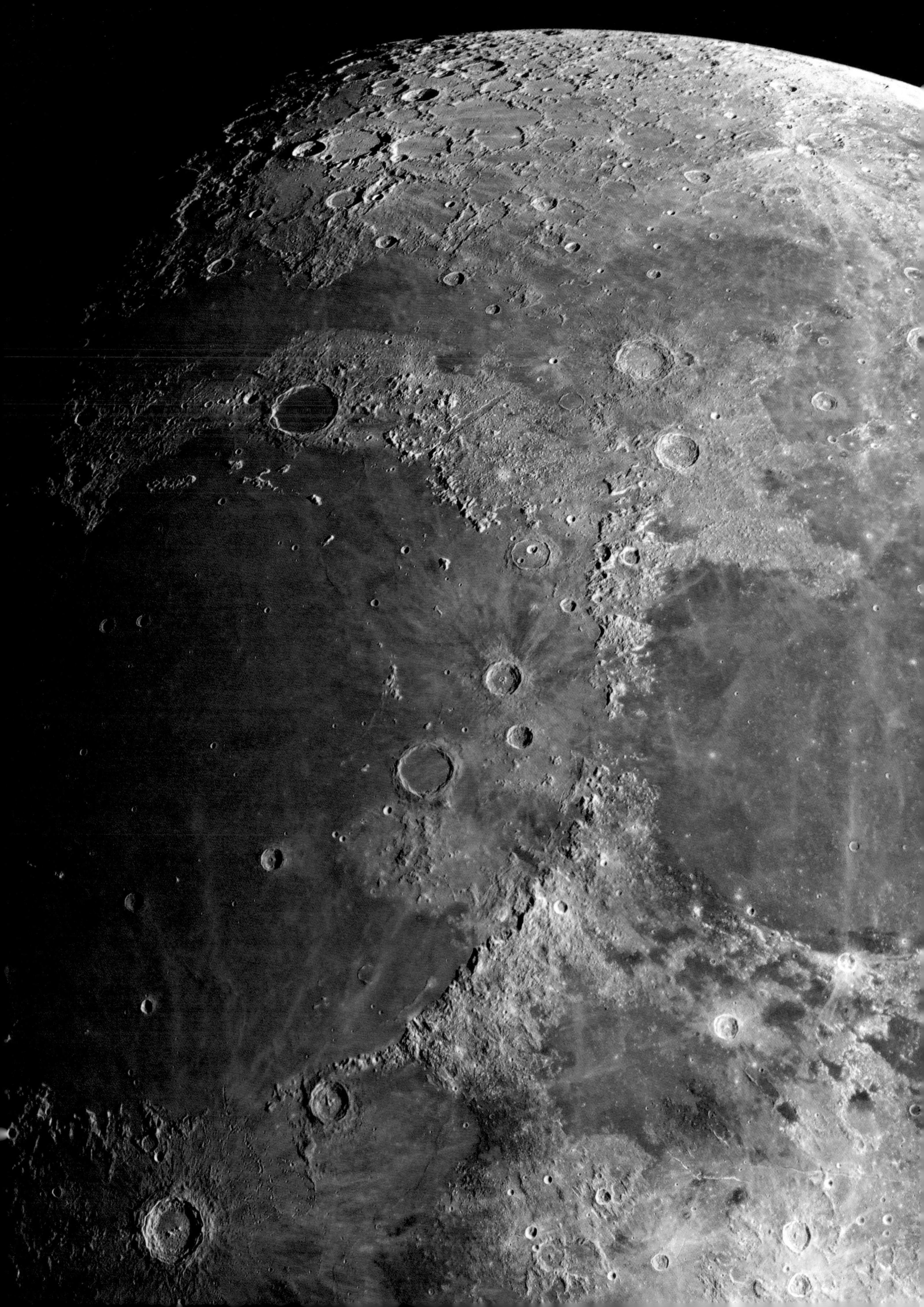

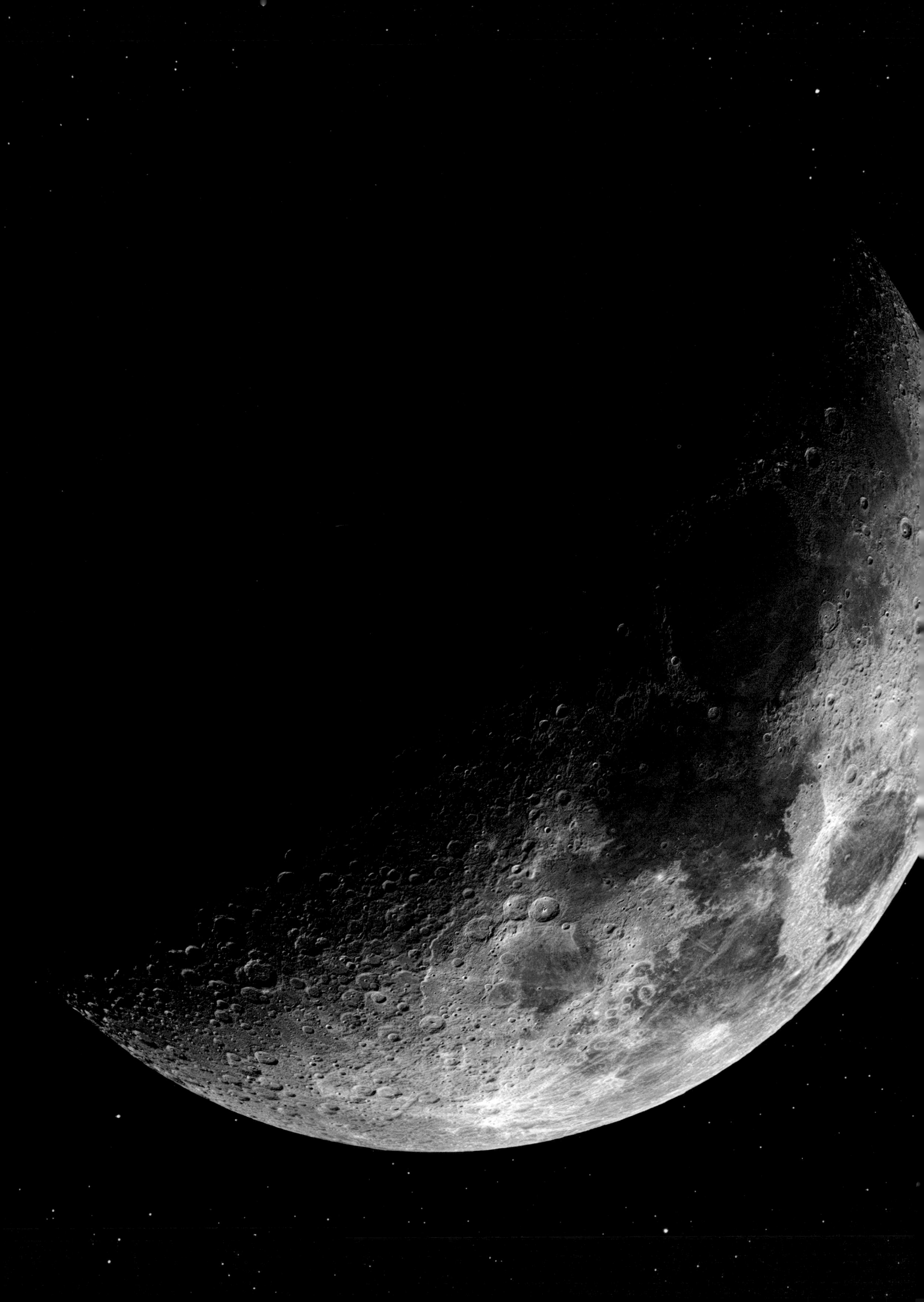

Waxing Crescent Moon

Capturing the Moon a little before First Quarter in the lunar cycle, we can see more detail in the parts just emerging from shadow at the terminator – the boundary between dark and light.

If you were standing in a spacesuit on the lunar surface at the terminator, you could see the Sun at the horizon, but since there is no atmosphere on the Moon the Sun would look every bit as brilliant and white as it does when it is overhead in the middle of the lunar day, and it would be set against a jet black sky. Shadows would be long and extremely dark. This extreme level of light/dark contrast in the Moon's airless environment is why we see the craters and lunar features more clearly near the terminator.

The sunlit portion of the Moon is so bright that even the best cameras cannot record both background stars and the bright lunar landscape in the same exposure. The stars in this image were exposed separately. All images in the composite from Canon EOS-20D / LX200.

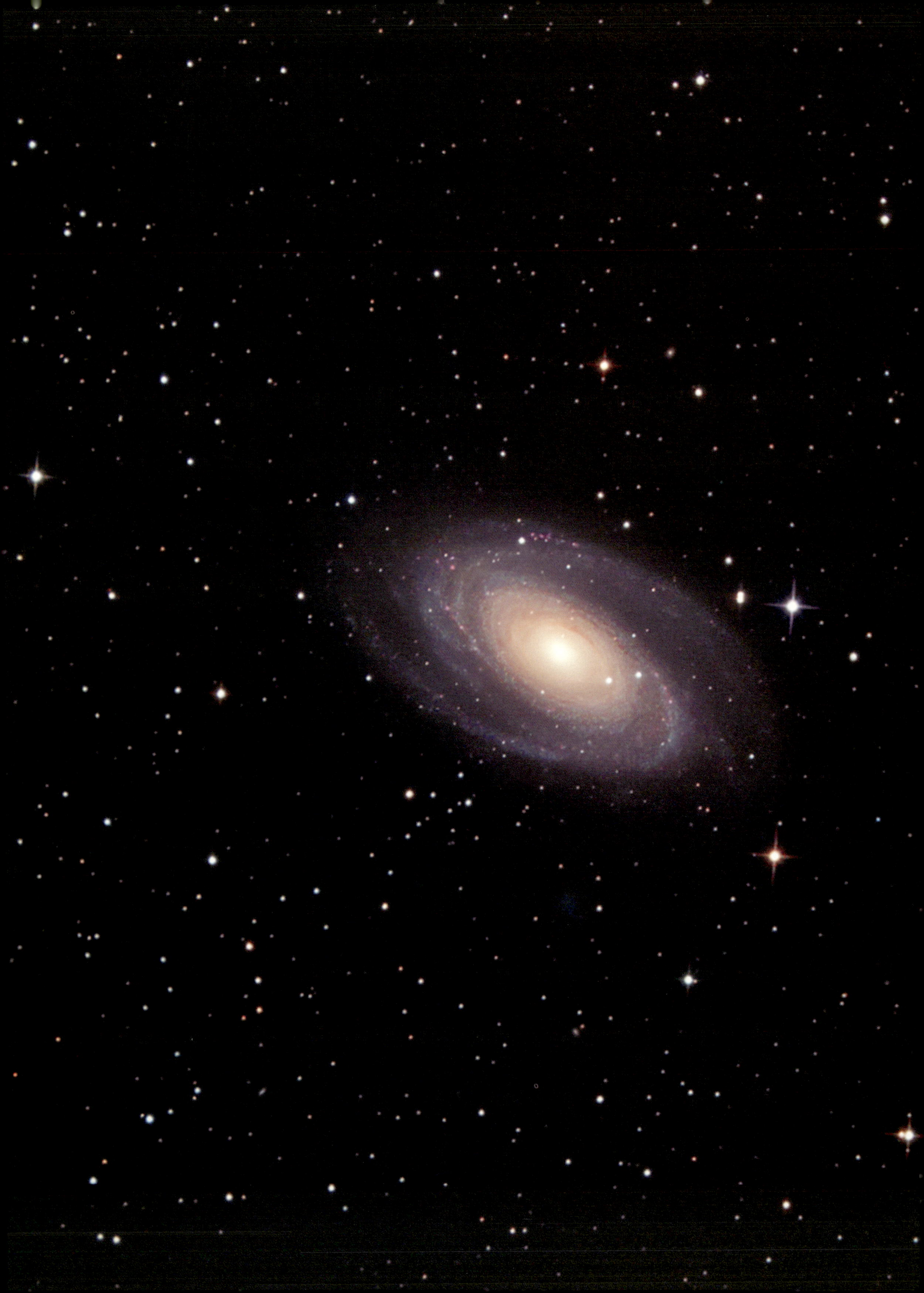

M81 – Spiral Galaxy

Once called Bode's Nebula, M81 is a beautifully symmetrical spiral galaxy in Ursa Major, near M82 (the Cigar Galaxy) and NGC3077, as seen in the wide-field image below.

Just beneath M81 is a small and very blue satellite galaxy called Holmberg IX, which shines brightly in ultraviolet light. It also contains one of only two known peanut-shaped yellow supergiant eclipsing binary stars, though they cannot be seen at this magnification.

In the deep exposure below we can begin to see traces of the Integrated Flux Nebula, which permeates the entire region.

M81 image (left): 2 hours 10 minutes RGB exposure, 3 hours 37 minutes Ha exposure, SXV-H9C / Hyperstar; wide-field image (below): 7 hours RGB, SXVF-M25C / Sky-90.

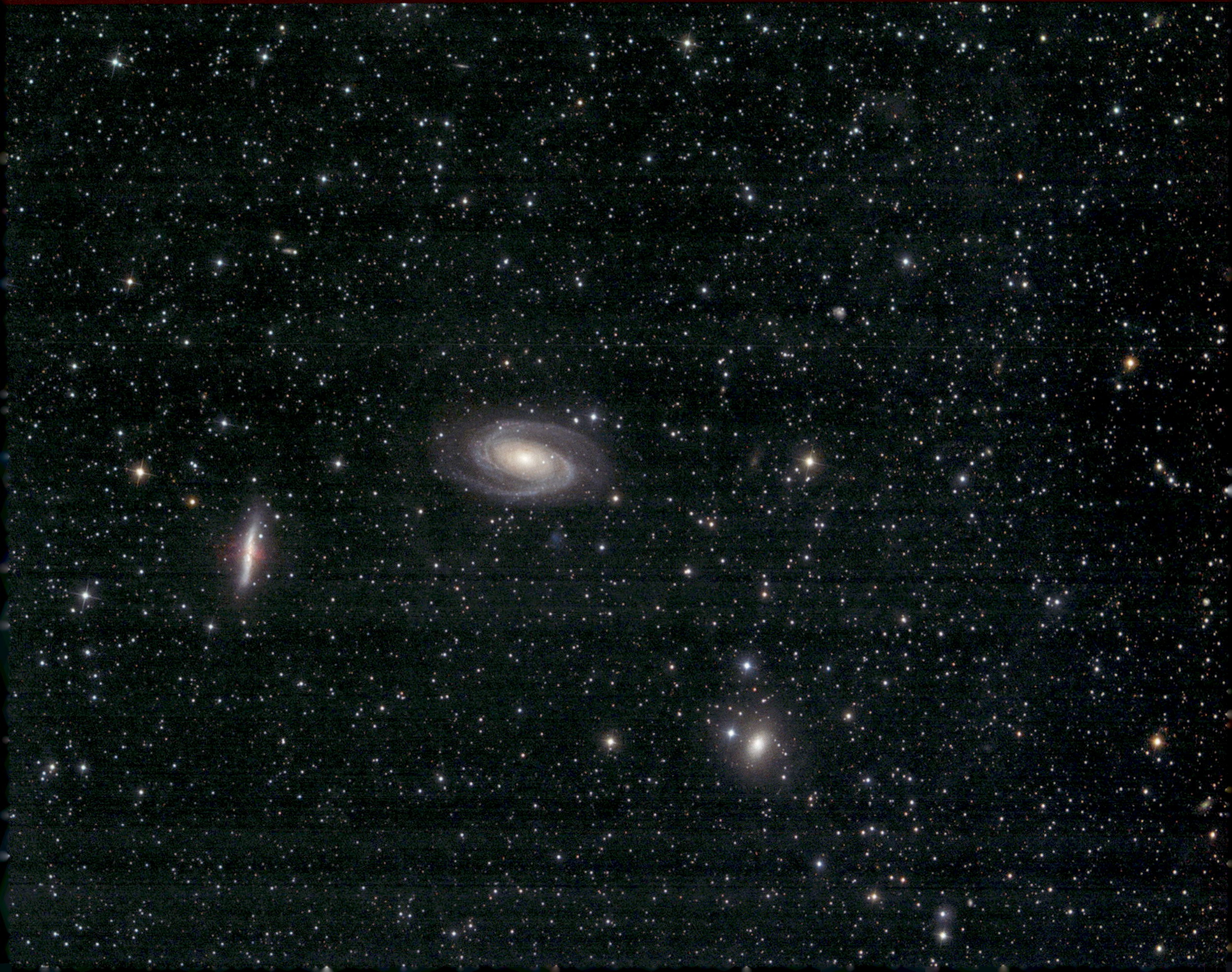

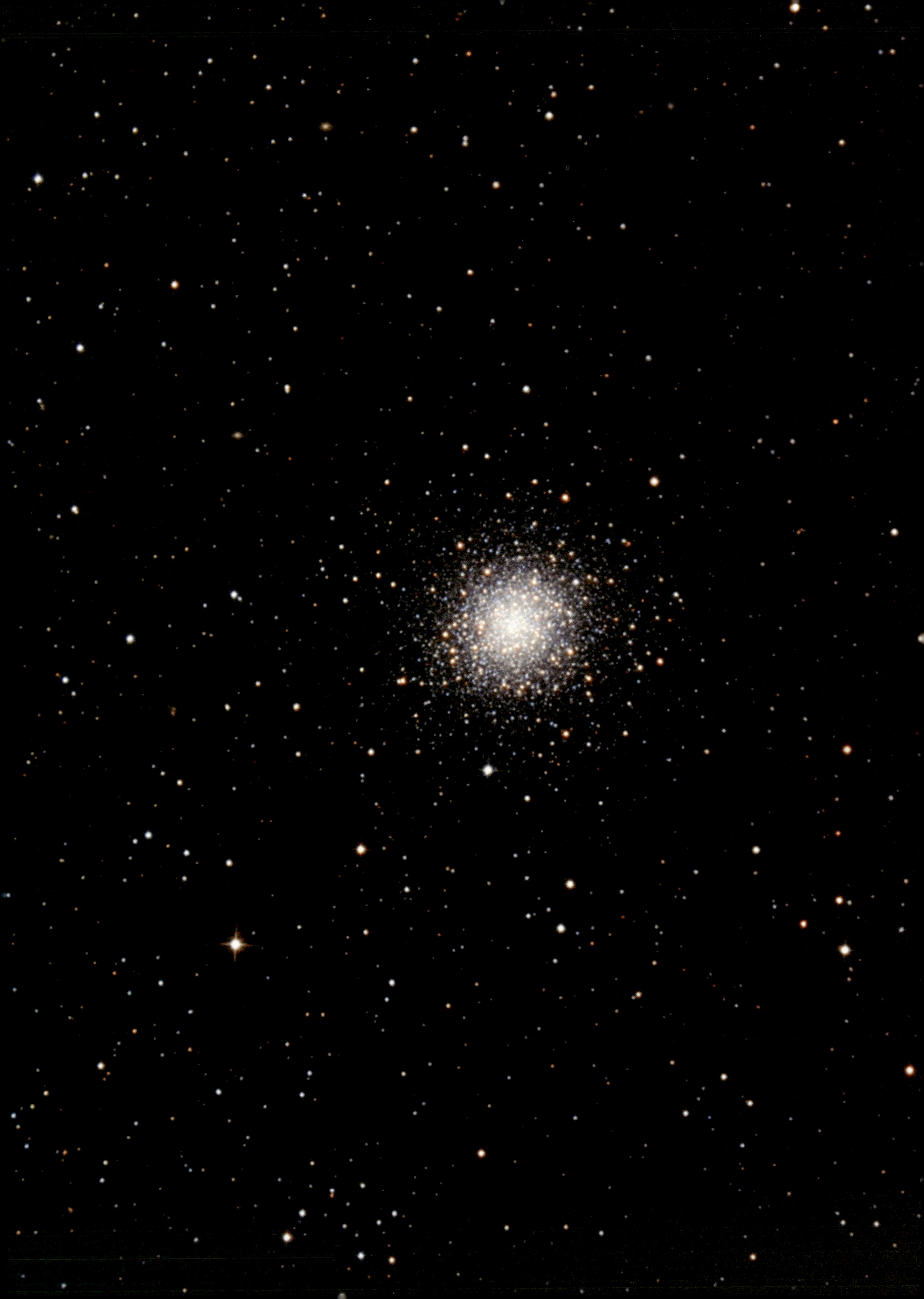

M92 – Globular Cluster in Hercules

A bright and beautiful globular cluster in the constellation Hercules, M92 is more compact than the more well known M13 (Hercules Cluster) nearby.

Even at 26,000 light-years distant, M92 can be visible to the naked eye under very dark conditions. This cluster is hurtling toward us at 250,000 miles per hour – but don't worry, it won't be here for another 71 million years!

The image below is a less deeply exposed and more highly magnified view of the cluster core; see if you can match up the stars. (Hint: The field of view is only about one quarter the width of the image at left.)

M92 image (left): 1 hour 9 minutes exposure, SXV-H9C / Hyperstar; core close-up image (below): 13 minute exposure, EOS-20D / LX200.

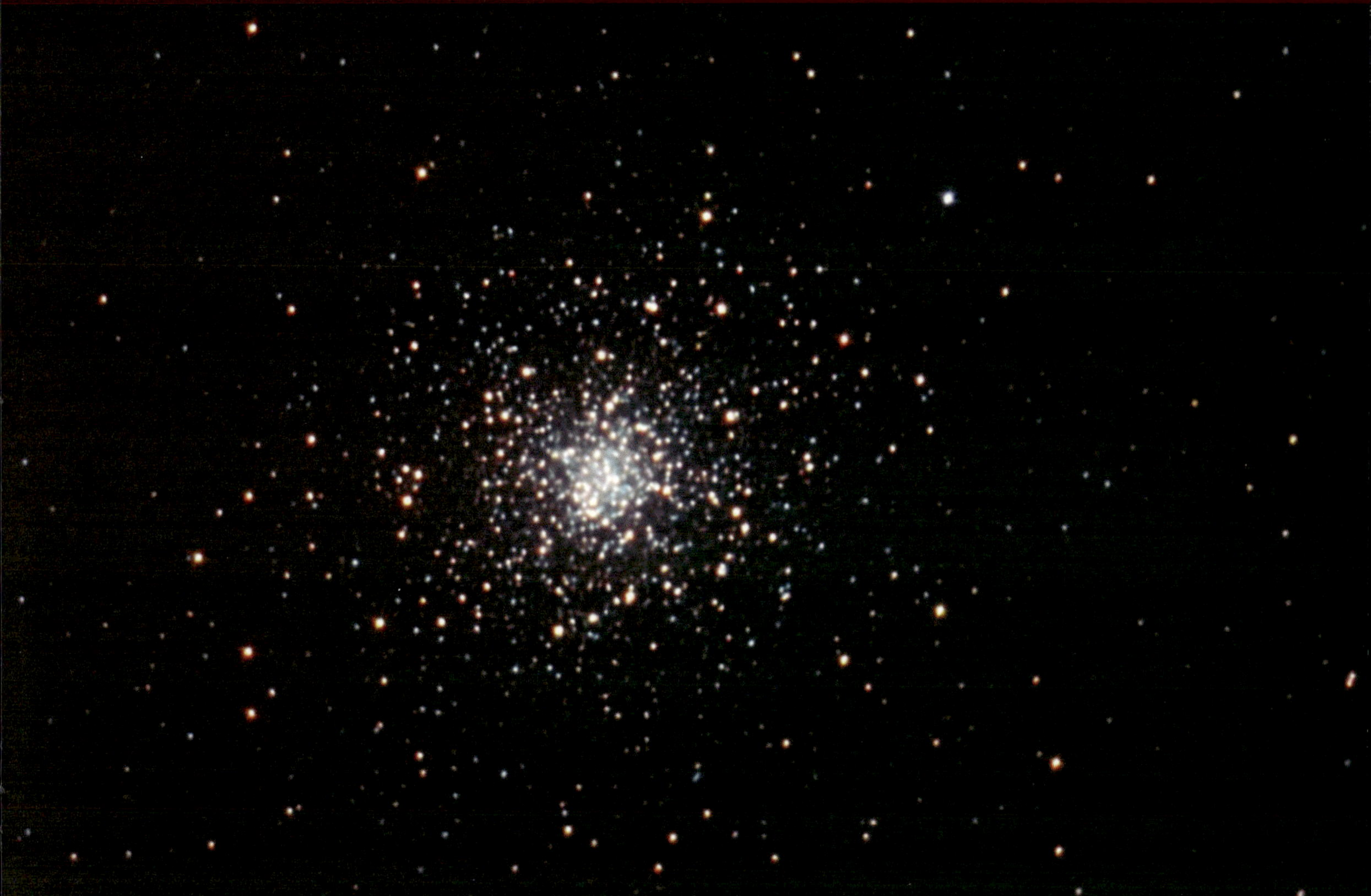

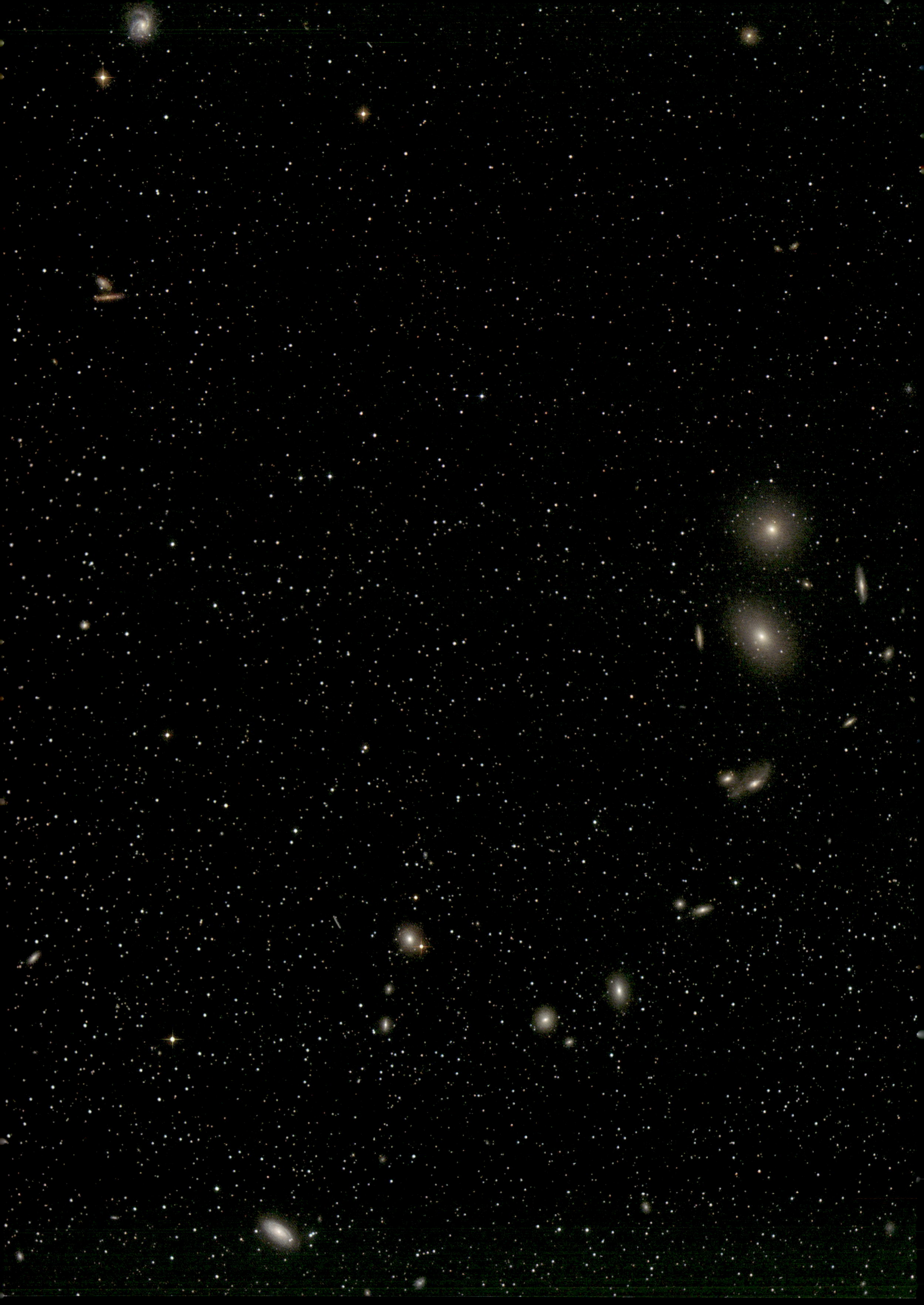

< Previous Two Pages

The Markarian Chain of Galaxies

This huge two-frame M25C/Sky-90 image of the Markarian Chain region gives a very wide field of view of 3 degrees by 3.5 degrees, while the diameter of the full Moon is only half a degree in size. It is presented in both positive and negative views to help show the positions of the galaxies.

The center of this cluster of galaxies lies some 70 million light-years away. Some extra galaxies to note are, at the top-left the beautiful spiral galaxy M99, towards the bottom-right M87, and towards bottom-left M88. There are three asteroids in this image, appearing as short line segments (they moved slightly during the exposures). Just left of the galaxy with the red star is 579 Sidonia, off to the right of M99 is 286 Iclea, and immediately below M87 is 599 Luisa.

Markarian Chain image: 8 hours total exposure, SXV-M25C / Sky-90

Opposite >

NGC869 and NGC884 – The Double Cluster in Perseus

The beautiful double (open) cluster in Perseus has been known s
antiquity. Lying some 7,300 light-years away, these objects are c
described by visual observers as looking like "diamonds on black velvet.
either binoculars or a low-power telescope these jewels are a wond
sight. With the naked eye the Double Cluster appears as a single glo
patch of light, somewhat like a very large globular cluster.

The diameter of these clusters is only about 18 minutes of arc, so this im
was taken with the small SXV-H9C one-shot color camera on the Sky
refractor at f#4.5, the smaller camera giving a smaller field of view,
therefore a larger apparent magnification.

Double cluster image: 5 hours 45 minutes RGB exposure, SXV-H9C / Sky

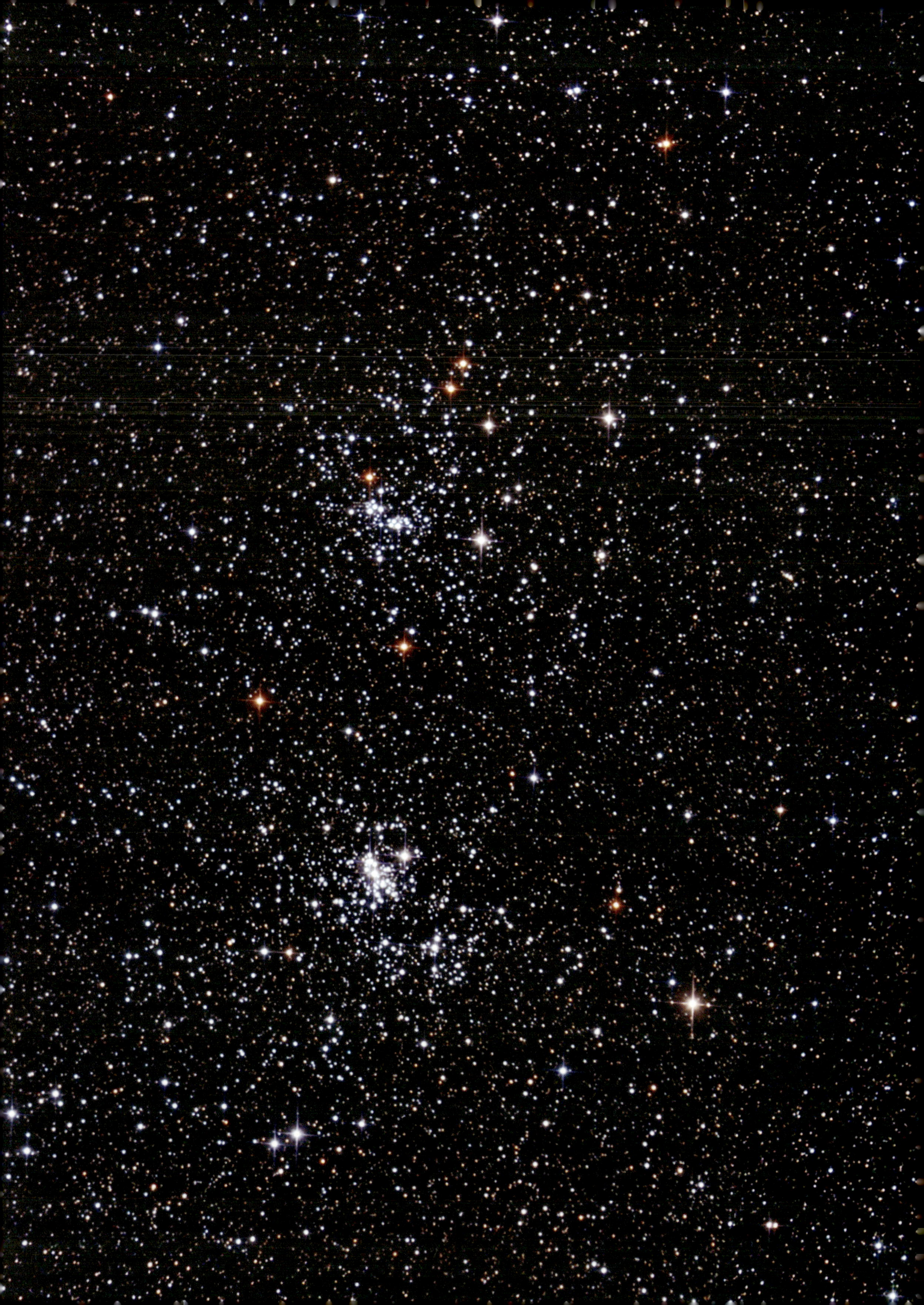

NGC6791 – Open Cluster in Lyra

Resembling a globular cluster because of its symmetry and large number of stars, this old open cluster has been discovered to be one of the most metal-rich clusters in the Milky Way, which in astrophysical terms means that the stars contain more elements heavier than helium than do the stars of other clusters. What makes this interesting is that it contradicts the general rule that older stars should be less metal-rich. Thus, NGC6791 is one of the most studied clusters in the sky.

NGC6791 image: 45-minute exposure, SXV-H9C / Hyperstar.

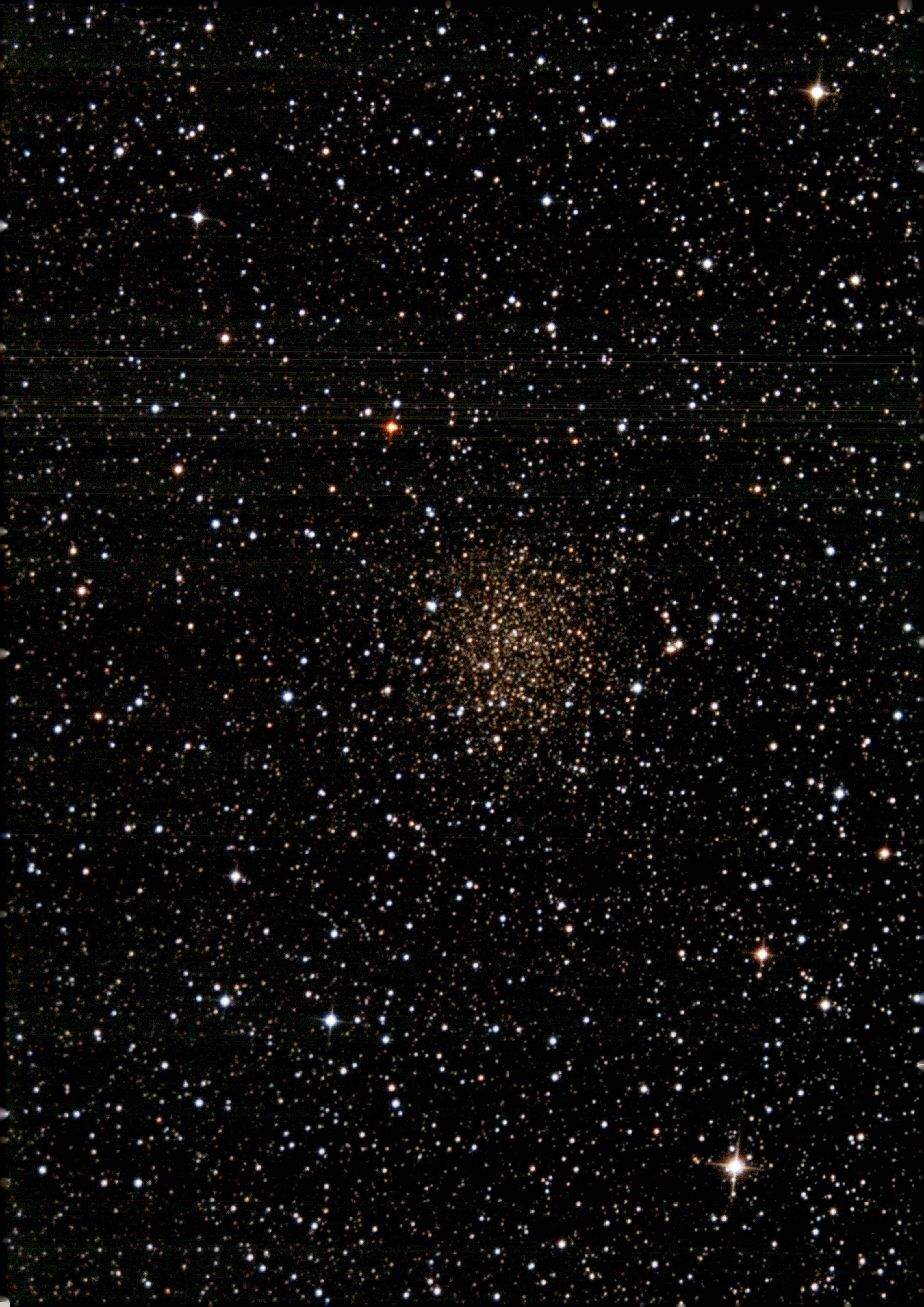

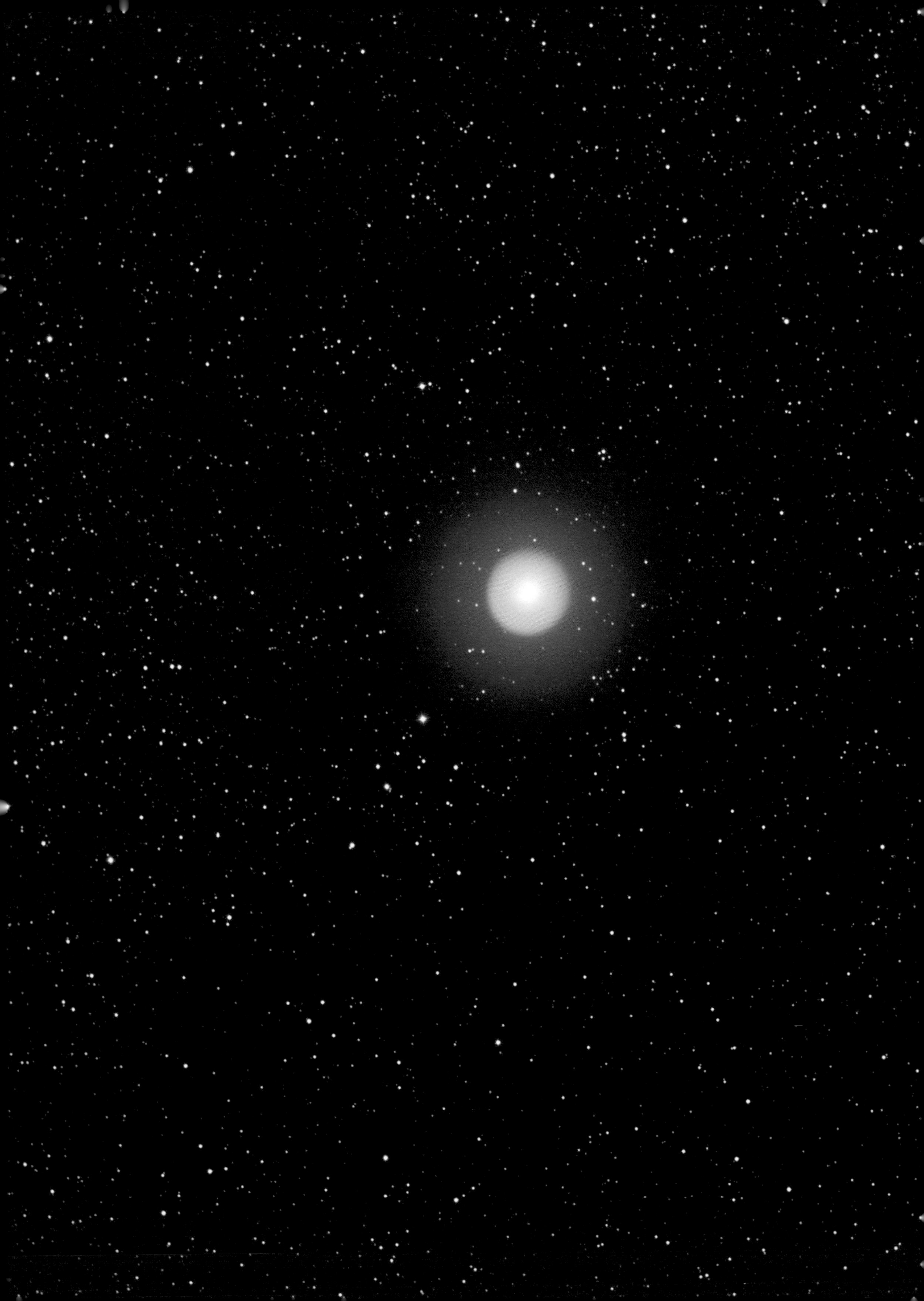

Comet Holmes – Unexpected Brightening in 2007

In October of 2007 this comet, which never gets closer to the Sun than well beyond the orbit of Mars, unexpectedly brightened from a tiny, dim dot at magnitude 17 to a brilliant disk at magnitude 2, easily visible in binoculars and telescopes. Some think it may have been impacted by another body. Its coma continued to increase in size until it had expanded well beyond the apparent size of the full Moon.

Seen here are images at two different times. The image on the left is actually more highly magnified than the wide-field image below, which was taken two weeks later. The comet's tail is pointed almost directly away from us.

The green glow is from carbon and cyanogen molecules fluorescing in the sunlight.

Comet Holmes (left): 2 hours 10 minute exposure SXVF-M25C / Sky-90; wide-field (below) 12 minutes exposure, EOS-20D / 100-400 zoom lens.

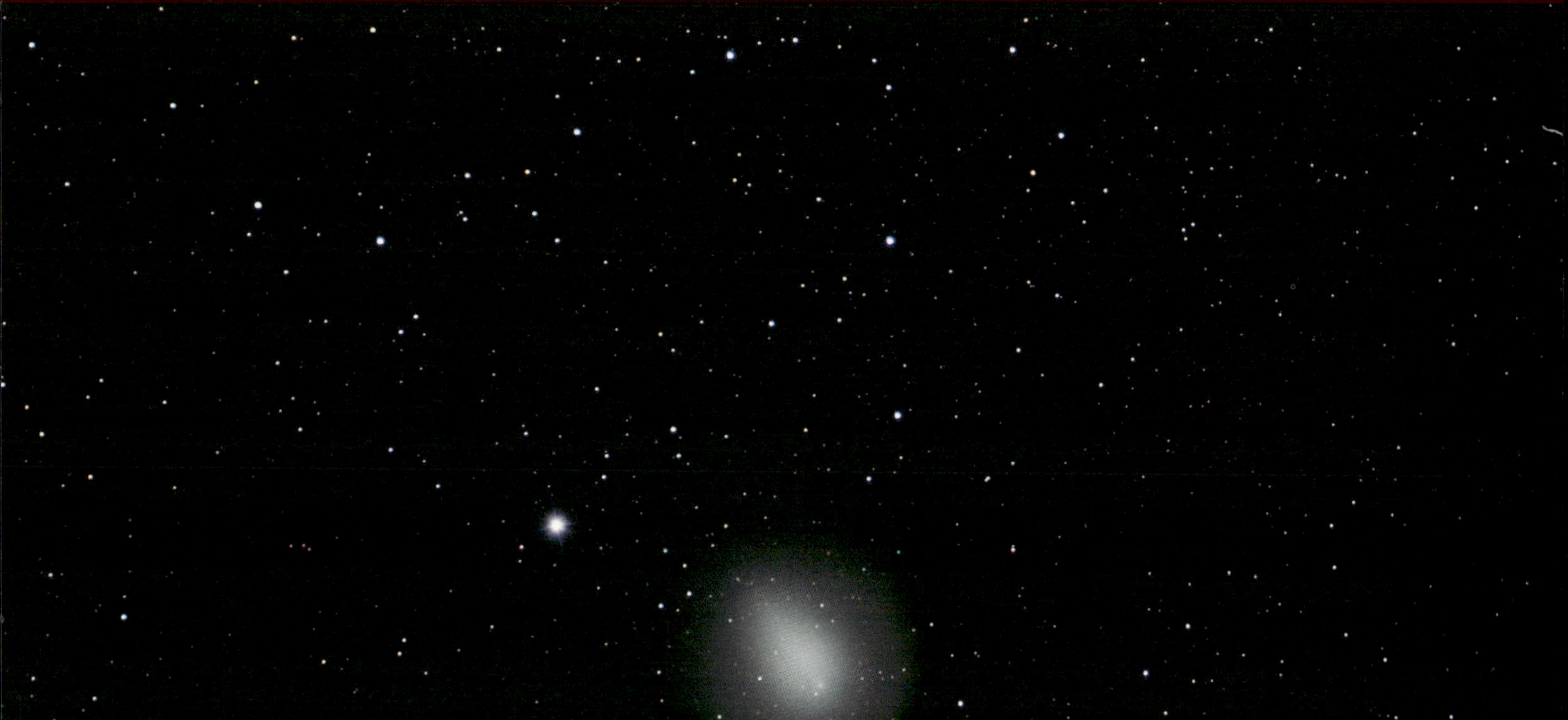

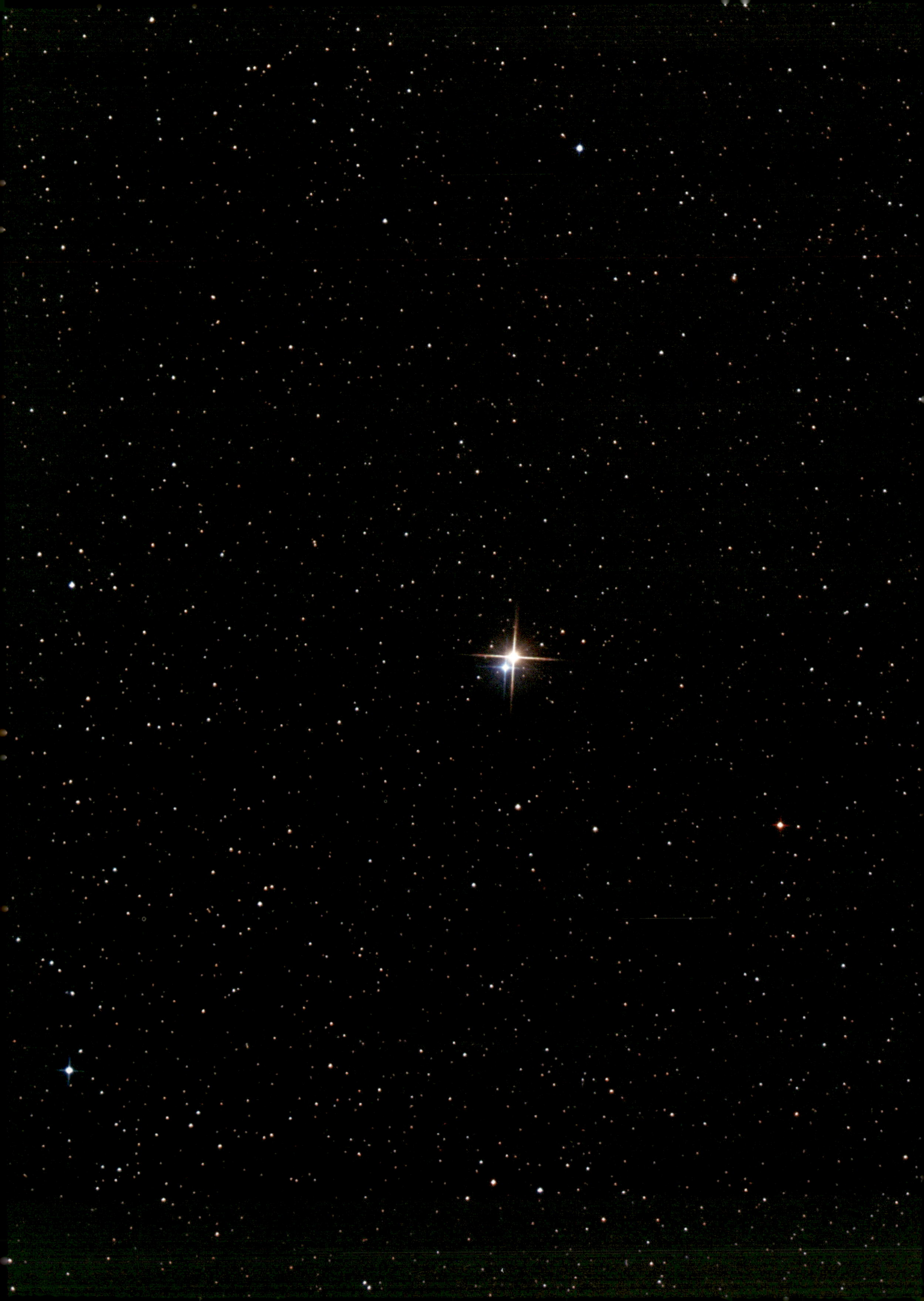

Albireo – A Colorful Double Star

At the base of the Cygnus cross (or the head of the Swan, depending how you want to look at it), what appears to be a single star to the naked eye under magnification resolves into a beautiful red/blue double. The two stars are truly paired, with one rotating around the other at a very great distance.

Not long ago what we see as the brighter, red star of this pair was actually found to be a very close double star itself. With the current separation between these a miniscule 0.4 seconds of arc, they are much too close together to be seen separately in this image but can be detected with very large telescopes.

Albireo image: 31 minutes exposure, SXV-H9C / Hyperstar.

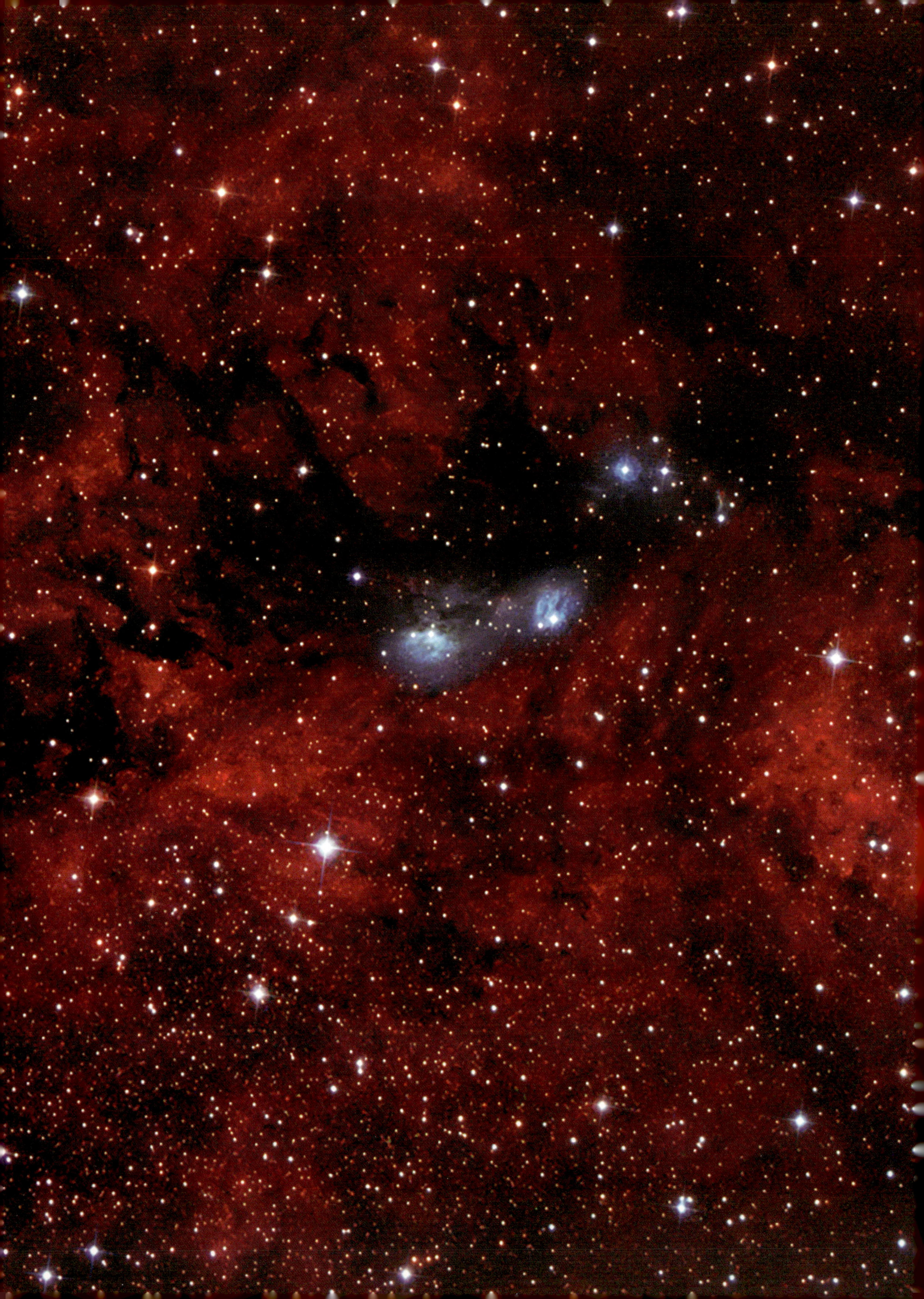

NGC6914 – Reflection Nebulae Among Emission Nebulae

NGC6914 identifies the blue reflection nebulosity seen in the center of this image, lying within a sea of red emissions that are part of the expansive Gamma Cygni nebula complex in central Cygnus. The reflection nebulae are cataloged in two parts as NGC6914A lying towards the south (to the right in this image), and NGC6914B lying towards the north.

The stars that illuminate NGC6914 belong to the Cygnus OB2 group, which is a massive grouping of young stars that may be the largest stellar association in our Milky Way.

NGC6914 image: 1 hour 51 minutes exposure, SXV-H9C / Hyperstar.

Leo Trio of Galaxies

Top to bottom the galaxies pictured are NGC3628, M65, and M66. Together in the constellation Leo they form the core of the M66 group and are commonly referred to simply as the Leo Trio, or the Leo Triplet.

These galaxies are probably gravitationally bound to one another. NGC3628 appears a bit jumbled with a dim trail of stars and showing signs of having been gravitationally disturbed, possibly by an astronomically recent merger with another galaxy.

At about 35 million light-years distant, these bright galaxies make for a beautiful photo and are also a nice visual challenge from a dark site.

Leo Trio image: Mosaic of two frames, 3 hours 37 minutes total exposure time, SXV-H9C / Hyperstar.

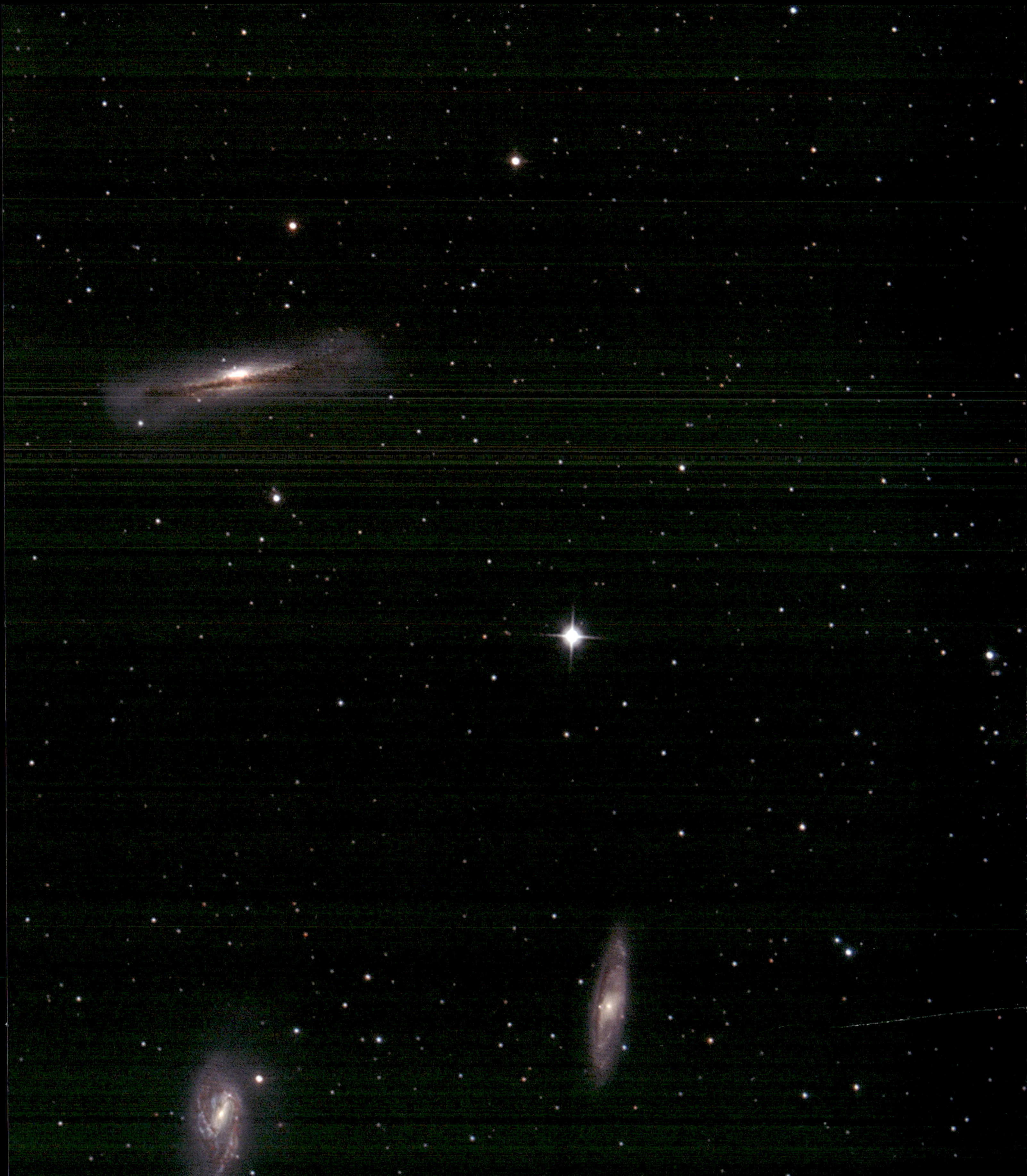

M67 – Open Cluster

An open cluster in Cancer, M67 lies at a distance of 2,600 light-years and contains roughly 500 stars, 100 of which are much like our own Sun.

M67 is an interesting open cluster due to its relatively great age of between 4 and 5 billion years – older than most open clusters.

M67 image: 3 hours 30 minutes exposure, SXVF-M25C / Sky-90. This image was our second joint APOD (Astronomy Photo of the Day), published on August 9, 2007.

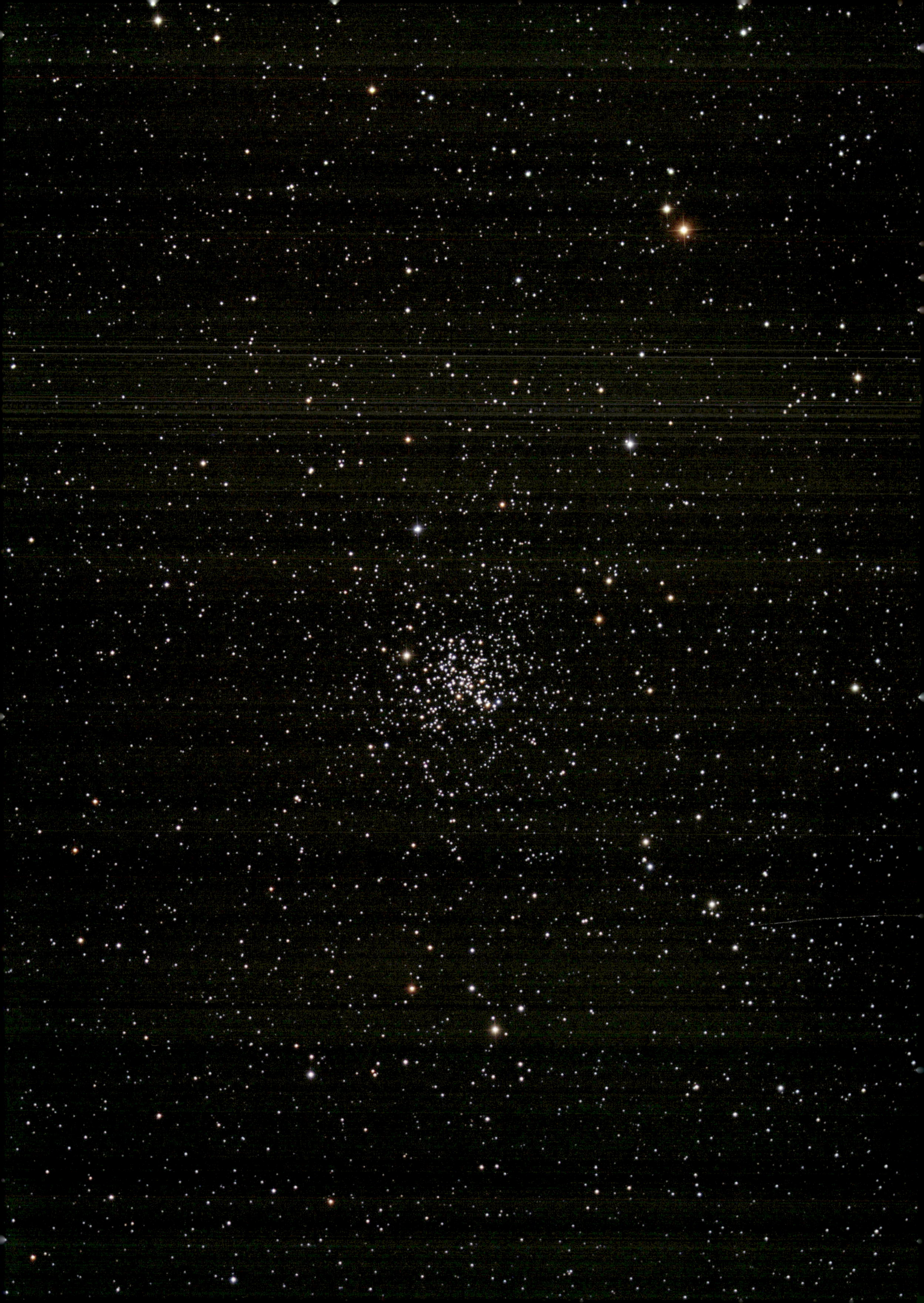

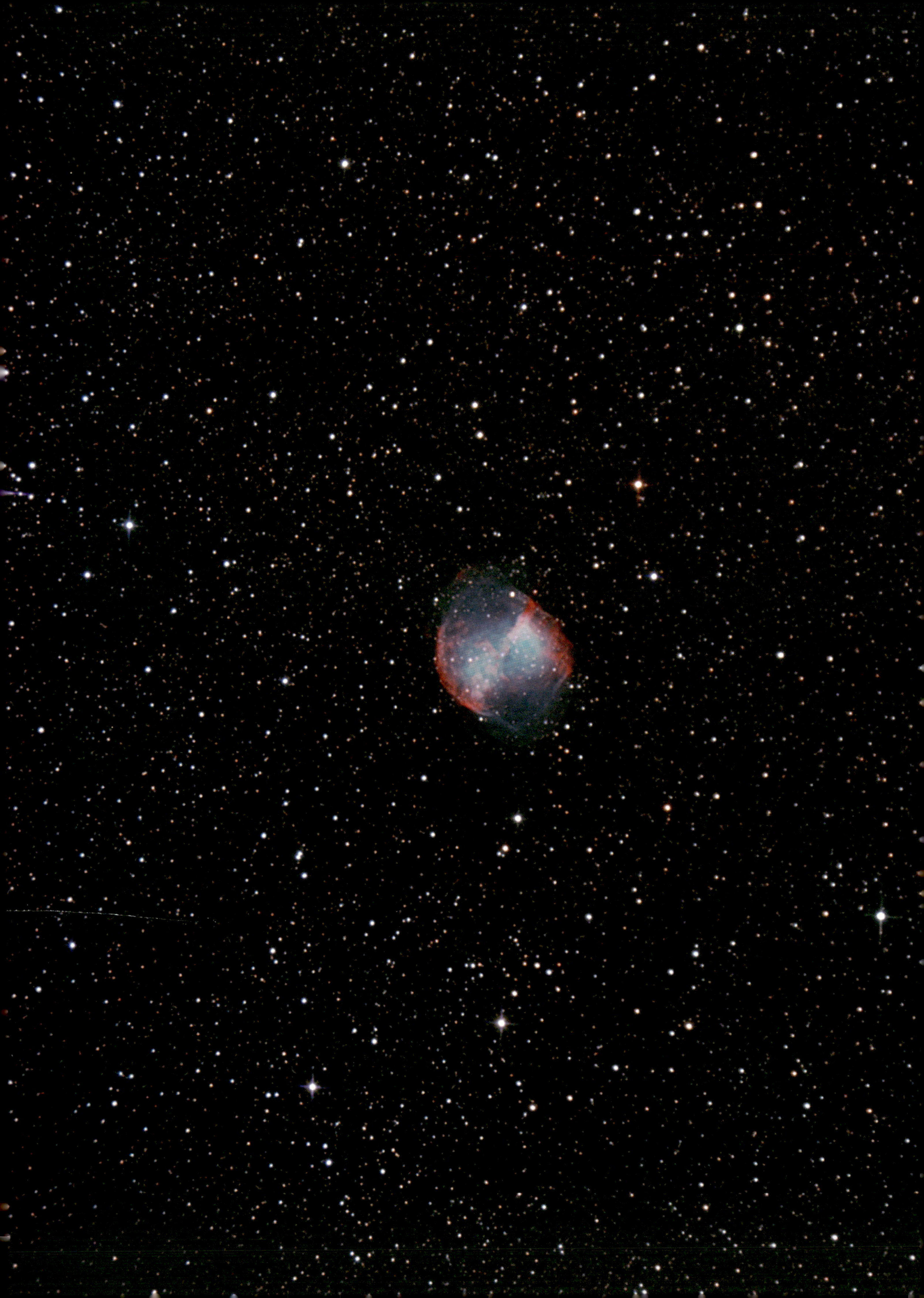

M27 – The Dumbbell Nebula

This planetary nebula is illuminated by its extremely hot blue-white central white dwarf star. It's estimated that the star lost its outer shell only 3,000 to 4,000 years ago and now has an amazingly high surface temperature of 85,000 Kelvin.

The reddish portions are made up of glowing hydrogen gas, while the bluish portions are made up of glowing oxygen. Is this the stuff of a future water world? A second home for humanity in the far distant future perhaps?

M27 image: 1 hour 31 minutes exposure, SXV-H9C / Hyperstar.

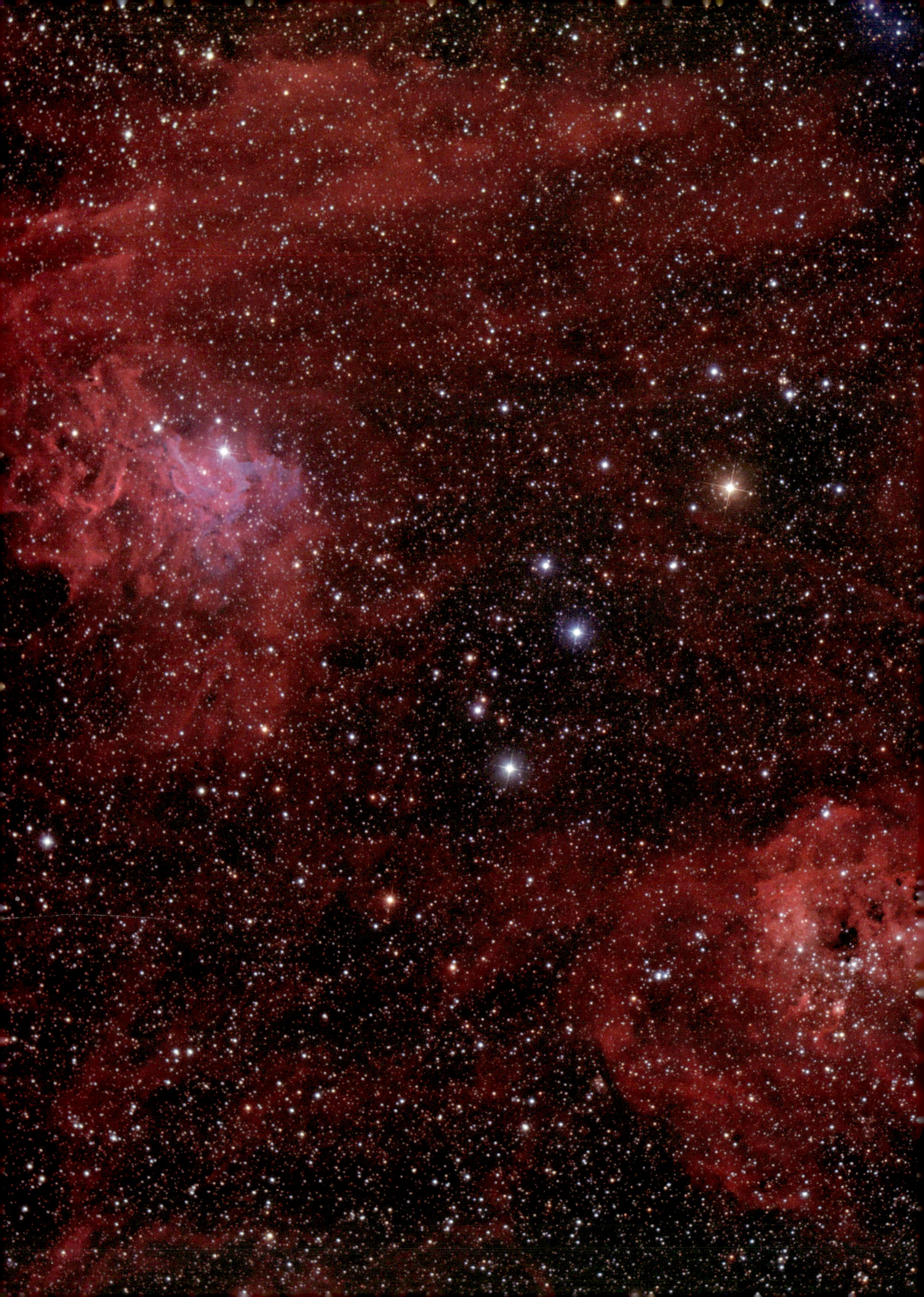

IC405 – Flaming Star Nebula and IC410

IC405, also known as the Flaming Star Nebula, is pictured at left and in the more magnified image below. IC410 is the bright nebula in the lower-right of the wide-field image at left.

These beautiful nebulae lie 1,600 light-years distant in the constellation Auriga. IC405 is particularly striking, with a pronounced blue reflection nebula adding to red emissions in the brightest region, yielding a violet coloring. Between the two nebulae is a colorful little open cluster of stars.

IC405 / IC410 image (left): 5 hours 37 minutes RGB and 6 hours Ha exposure, SXV-M25C / Sky-90; close-up (below) 1 hour 40 minutes exposure, SXV-H9C / Hyperstar.

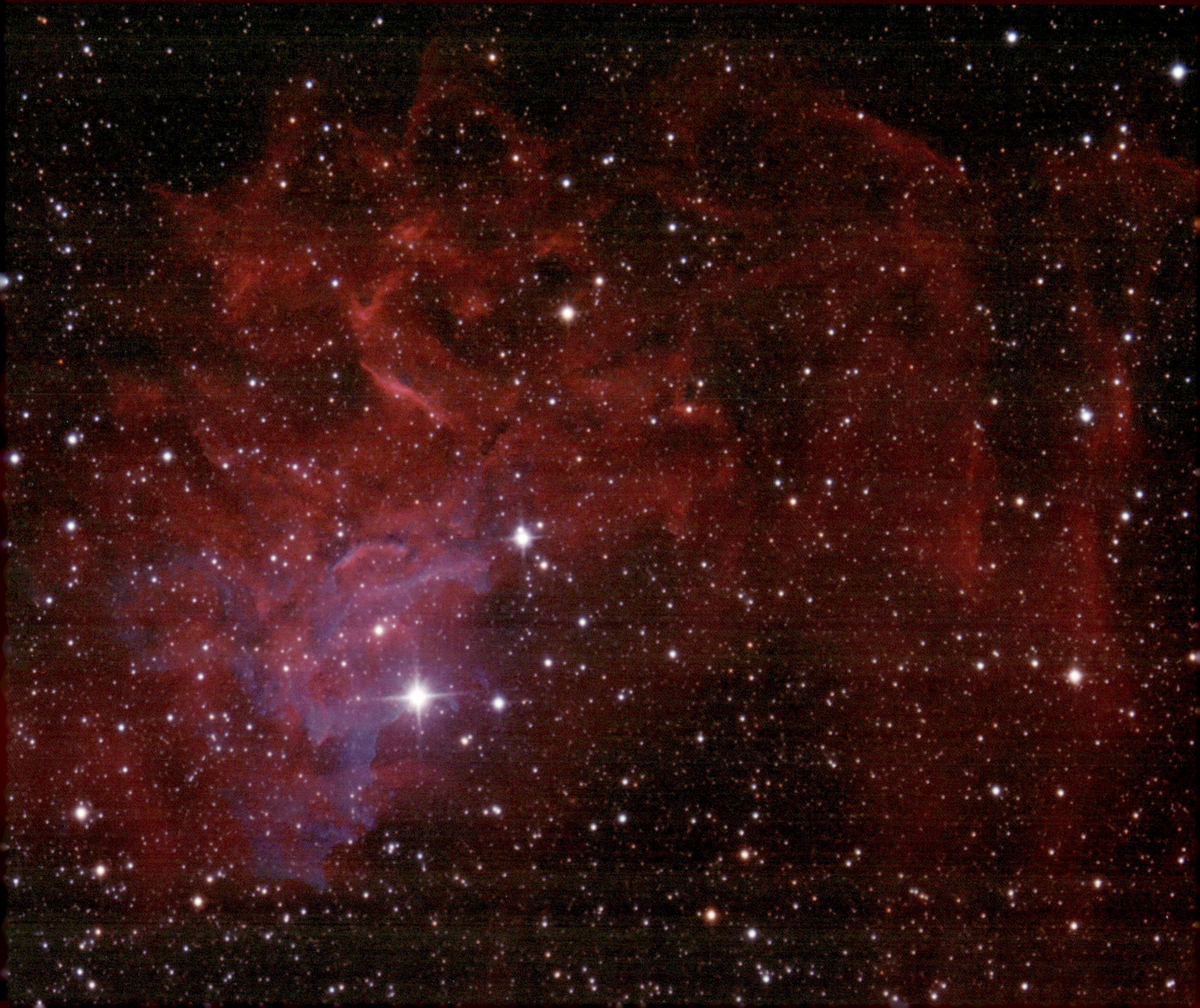

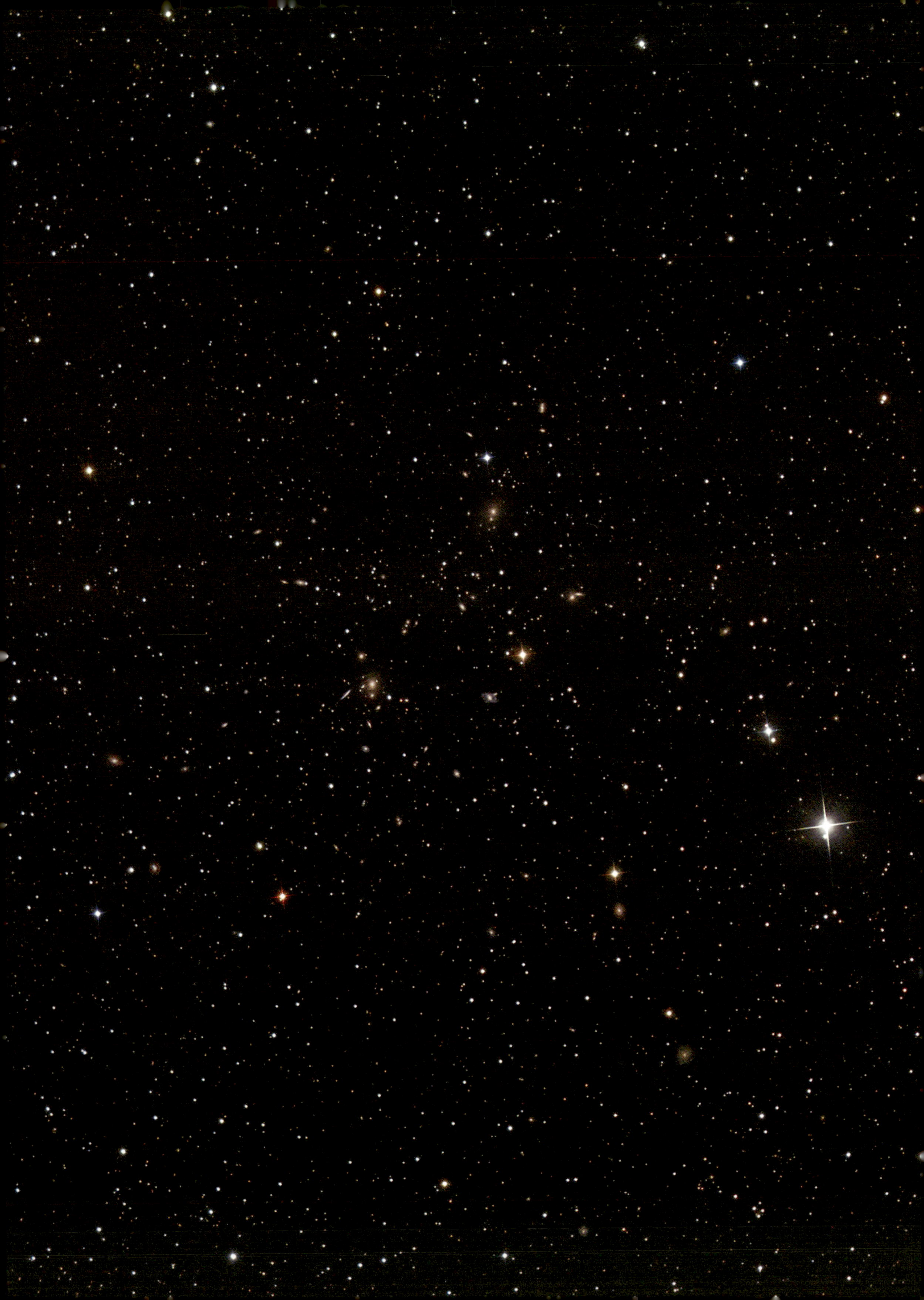

Abell 1367 is a fine cluster of galaxies in Leo, with more members brighter than magnitude 14 than in any other cluster. Abell 1367 lies at a distance of 330 million light-years, about the same distance as Abell 1656, and spans only about half a degree. Abell 1367 and Abell 1656 are the two largest clusters in the vast Coma super cluster of galaxies. Although the image scale of the M25C/Sky-90 is not ideal for imaging individual small objects like these distant galaxies, we capture an impressive number of "faint-fuzzies" here in the large 3.33 by 2.22 degree field of view.

Abell 1367 image: 4 hours 45 minutes exposure, SXVF-M25C / Sky-90.

Sadr and Surrounding Nebula Complex in Cygnus

The Gamma Cygni nebulosity surrounding the bright star Sadr is a huge collection of emission nebulae found in Cygnus. At magnitude 2.2, Sadr is at the center of the Cygnus Cross, also known as the Northern Cross.

The brighter portions of nebulosity toward the bottom of this image are sometimes referred to as the Butterfly Nebula, and a small open cluster of stars designated NGC6910 can be seen at the left-center.

Sadr image: 6 hours RGB exposure and 3 hours Ha exposure, SXVF-M25C / Sky-90 and including some additional earlier SXV-H9C/Hyperstar exposure data. This image was our first joint APOD (Astronomy Photo of the Day), published on January 7, 2007

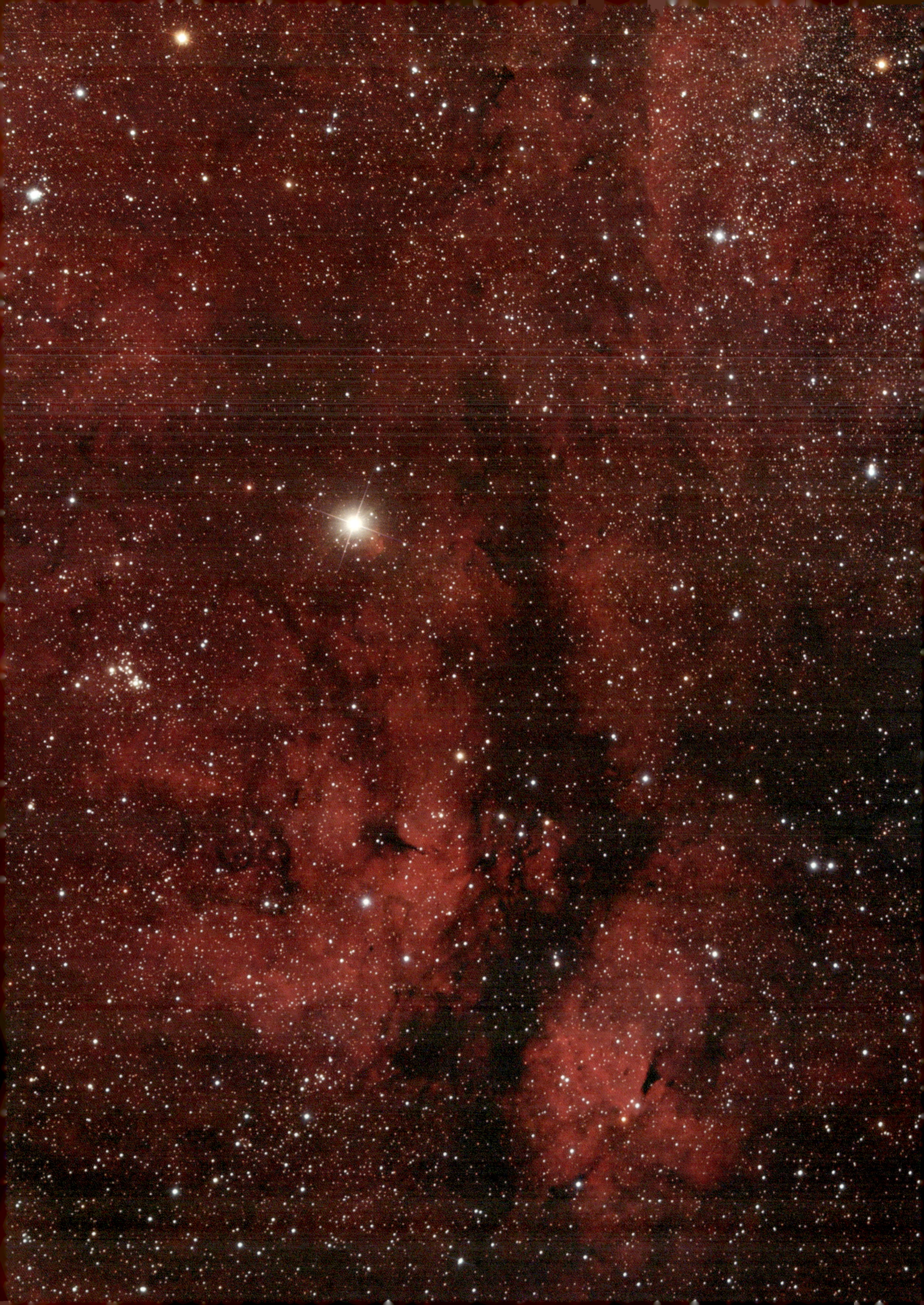

M35 and NGC2158 – Open Clusters

Turning the Sky90 towards the M35 region, we were surprised to see how much of the field of view was taken up by M35 alone. Our earlier close-up Hyperstar image of M35 (below) showed the cluster to be large, but we hadn't appreciated the full extent of this magnificent, bright blue open cluster of stars.

Nearby and contrasting in color we see the yellowish NGC2158, probably colored as such because it is shining through galactic dust, as it is about four times further off than M35's 2,700 light-years from Earth.

M35/NGC2158 close-up (below): 52 minutes exposure, SXV-H9C / Hyperstar; wide-field (right) 3 hours 54 minutes, SXVF-M25C / Sky-90.

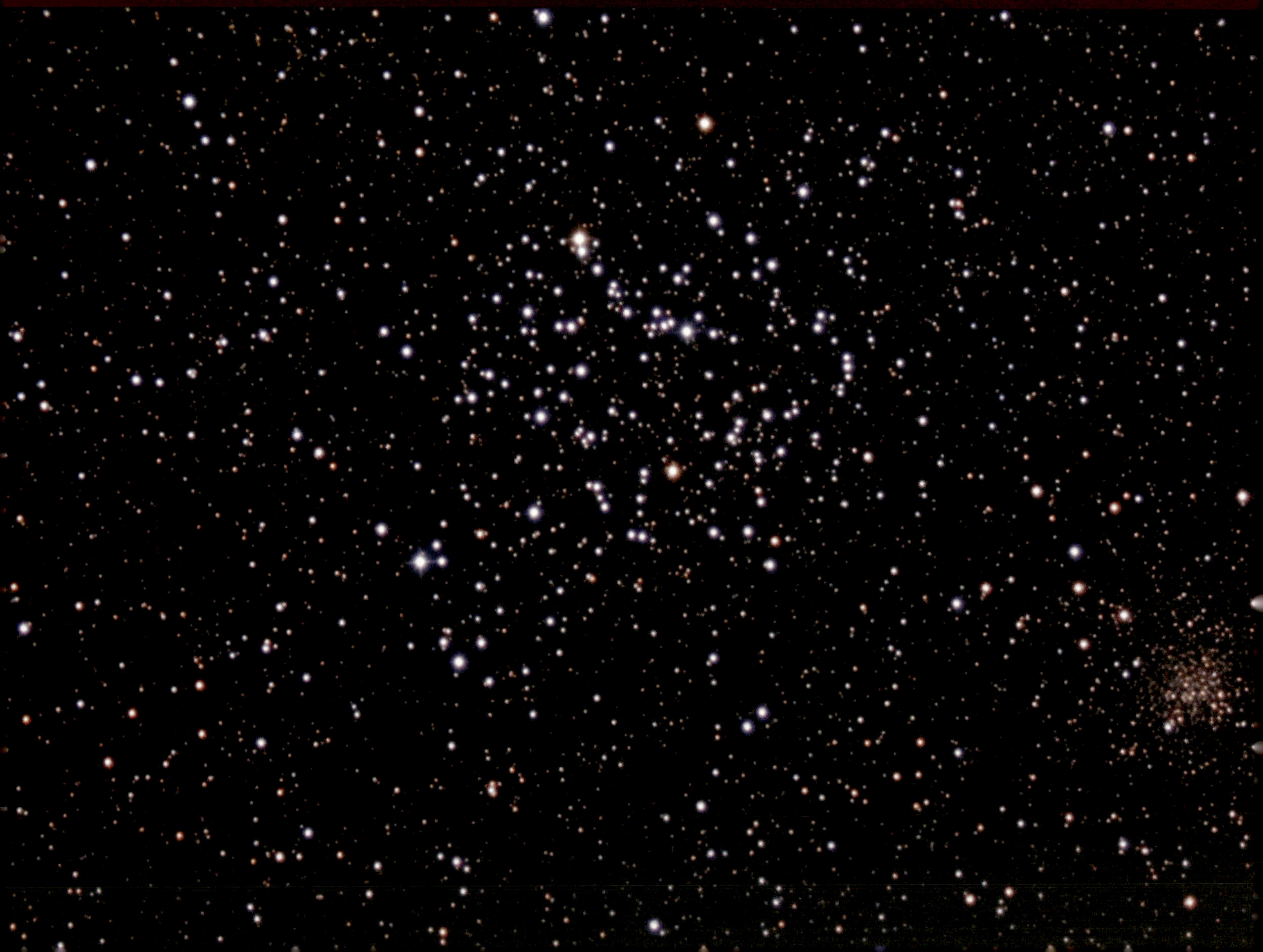

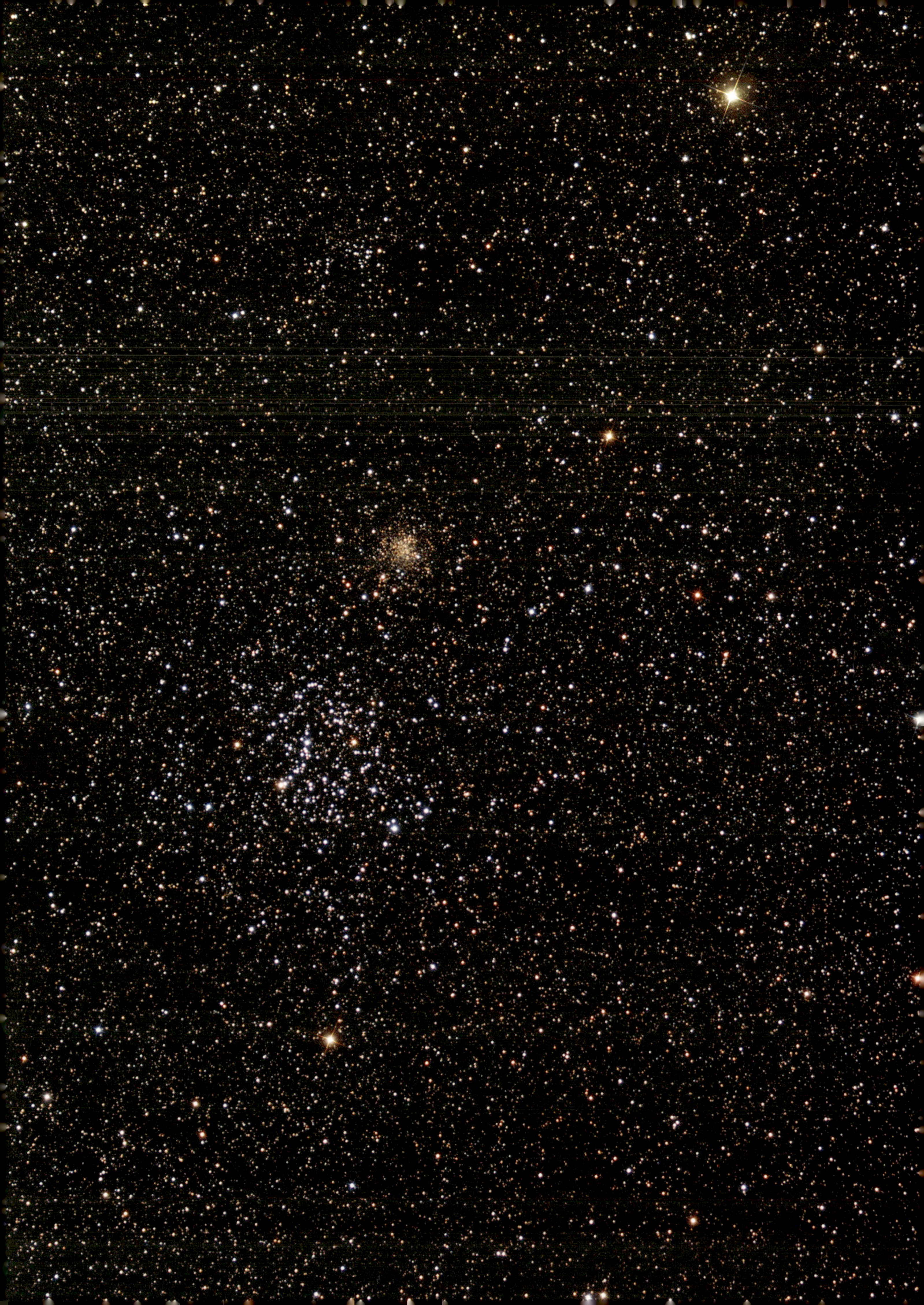

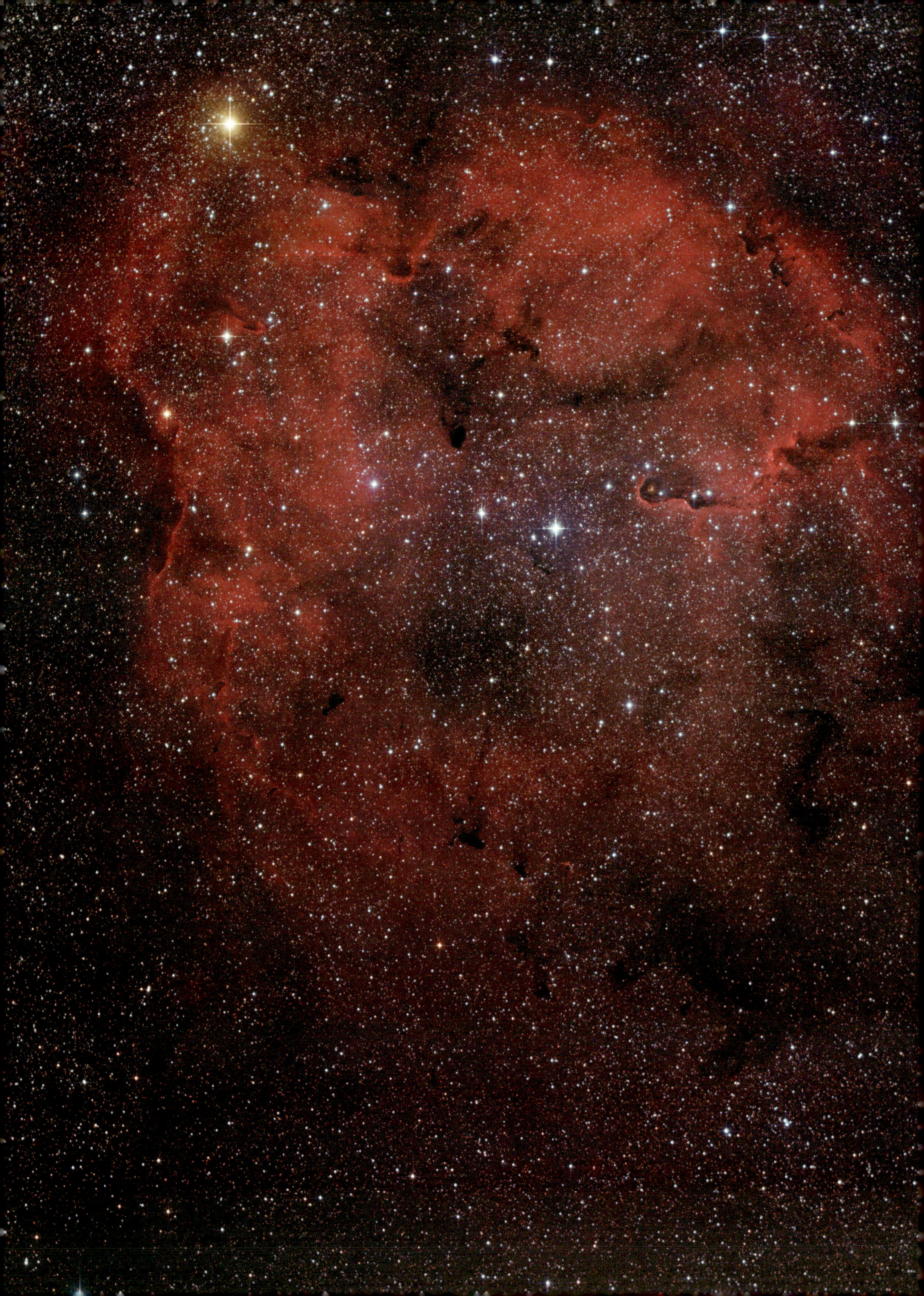

IC1396 – Featuring the Elephant's Trunk

There being no popular name for the overall nebula, this expansive area of emission and absorption nebulosity is simply known as IC1396 and lies 2,450 light-years from Earth in the constellation Cepheus.

Since it spreads over more than 3 degrees across the sky, this image is a 2-frame mosaic using the M25C/Sky-90 combination, giving an overall field of view approximately 3.33 by 4 degrees. The bright orange star at top-left is Mu Cephei, known as the Garnet Star, and the worm-like structure coming in from the right and ending in a loop is known as the Elephant's Trunk (shown magnified below).

IC1396 image: Mosaic of several frames, 7 hours 22 minutes total RGB, 11 hours 40 minutes total Ha exposure, SXVF-M25C / Sky-90.

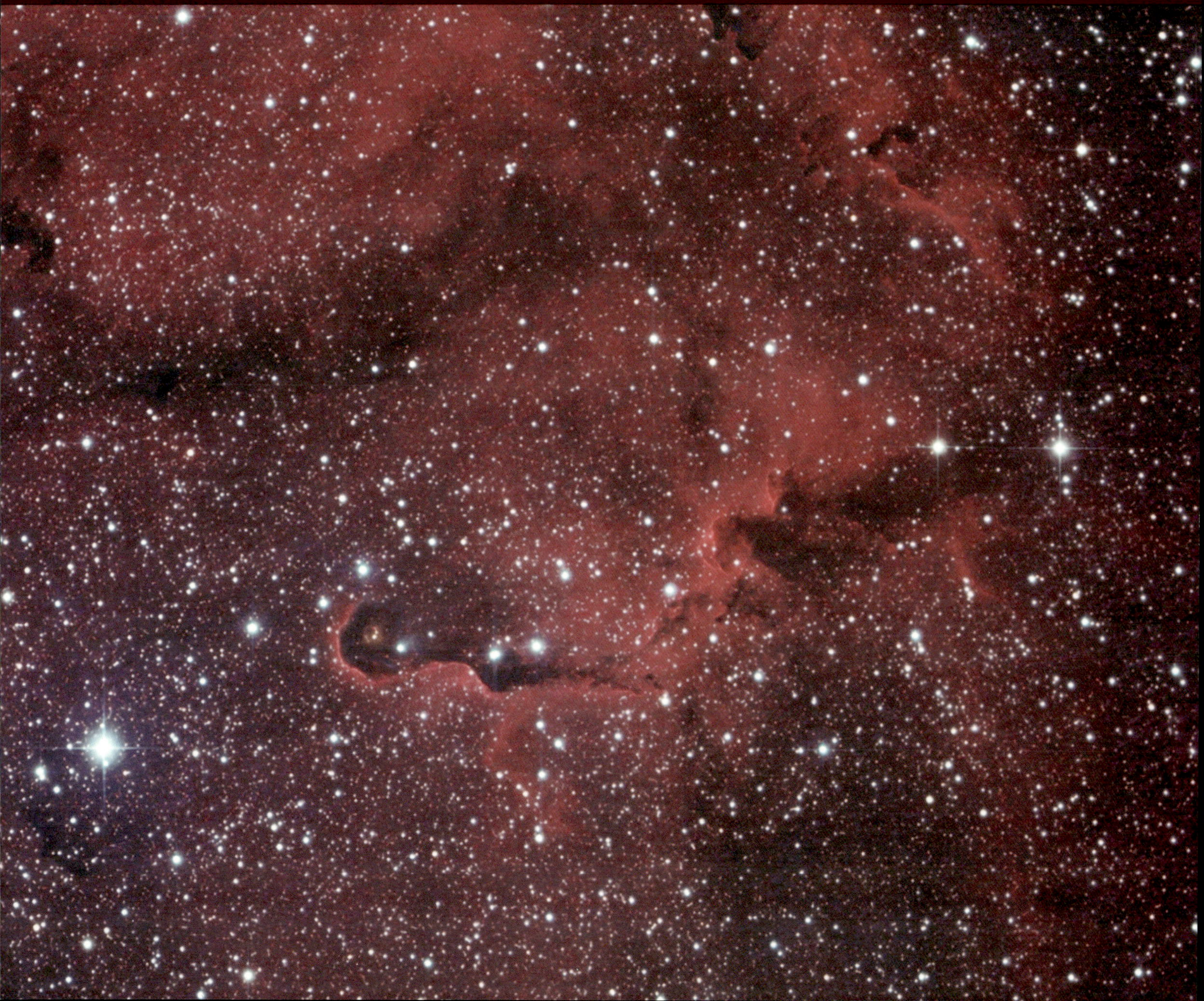

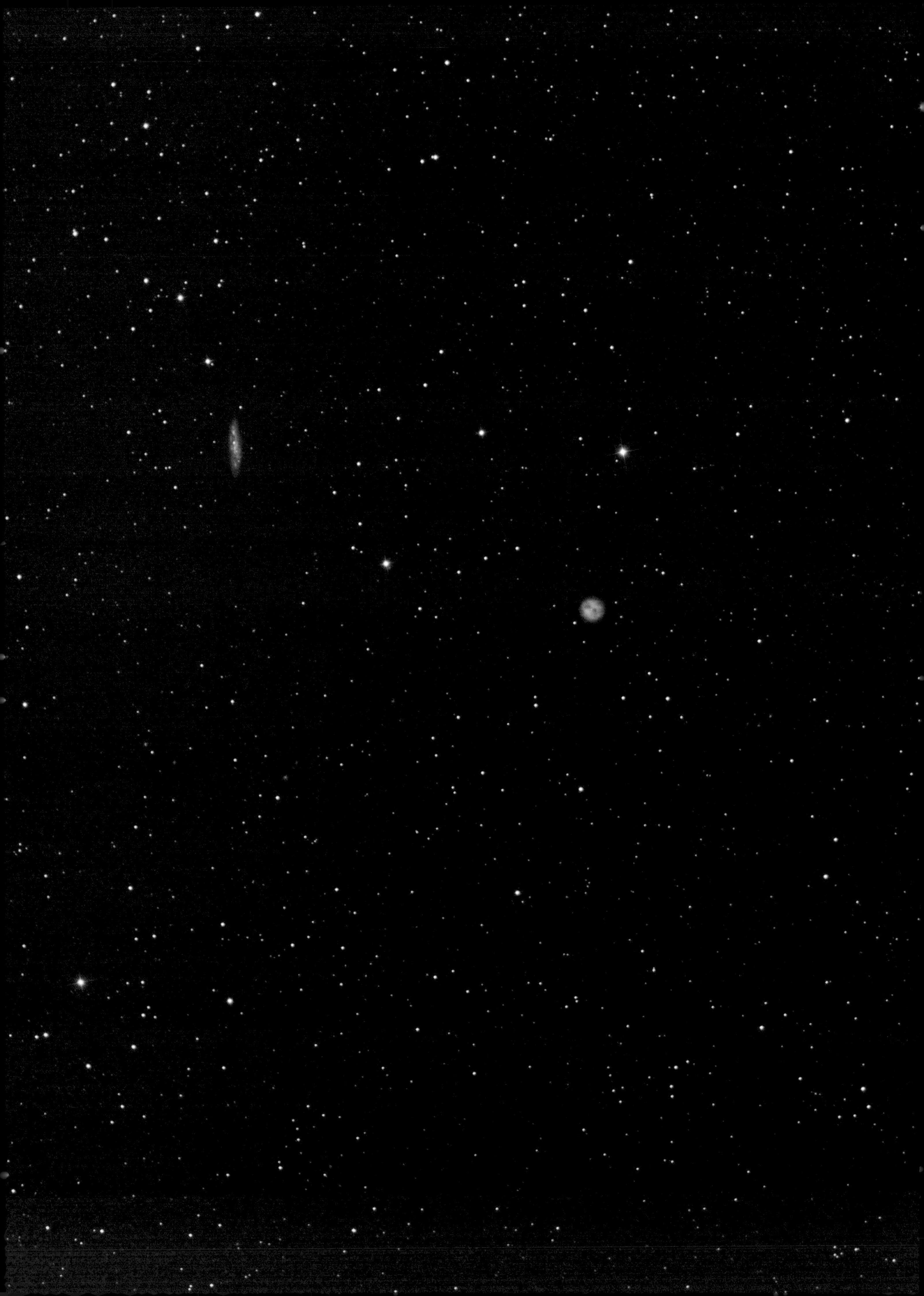

M97 – Owl Nebula and M108 Galaxy

Even though this is a very large field of view for studying such small objects, we can actually see some detail in M97, the Owl Nebula, and in the long, thin M108 galaxy in this part of the Ursa Major constellation.

The Owl Nebula is a mere 170 arc seconds across and lies 1,600 light-years from Earth. In this image you can just make out the two dark "eyes" that give this planetary nebula its name.

A tiny spiral galaxy seen almost edge-on, M108 seems similar in size to the nebula but is really a lot further off – a whopping 46 *million* light-years distant. Discovered in 1781 by Pierre Mechain, M108 is only about one-twentieth the mass of the Great Andromeda Galaxy, so it actually is quite small on celestial scales.

M97 / M108 image: 3 hours 45 minutes exposure, SXVF-M25C / Sky-90.

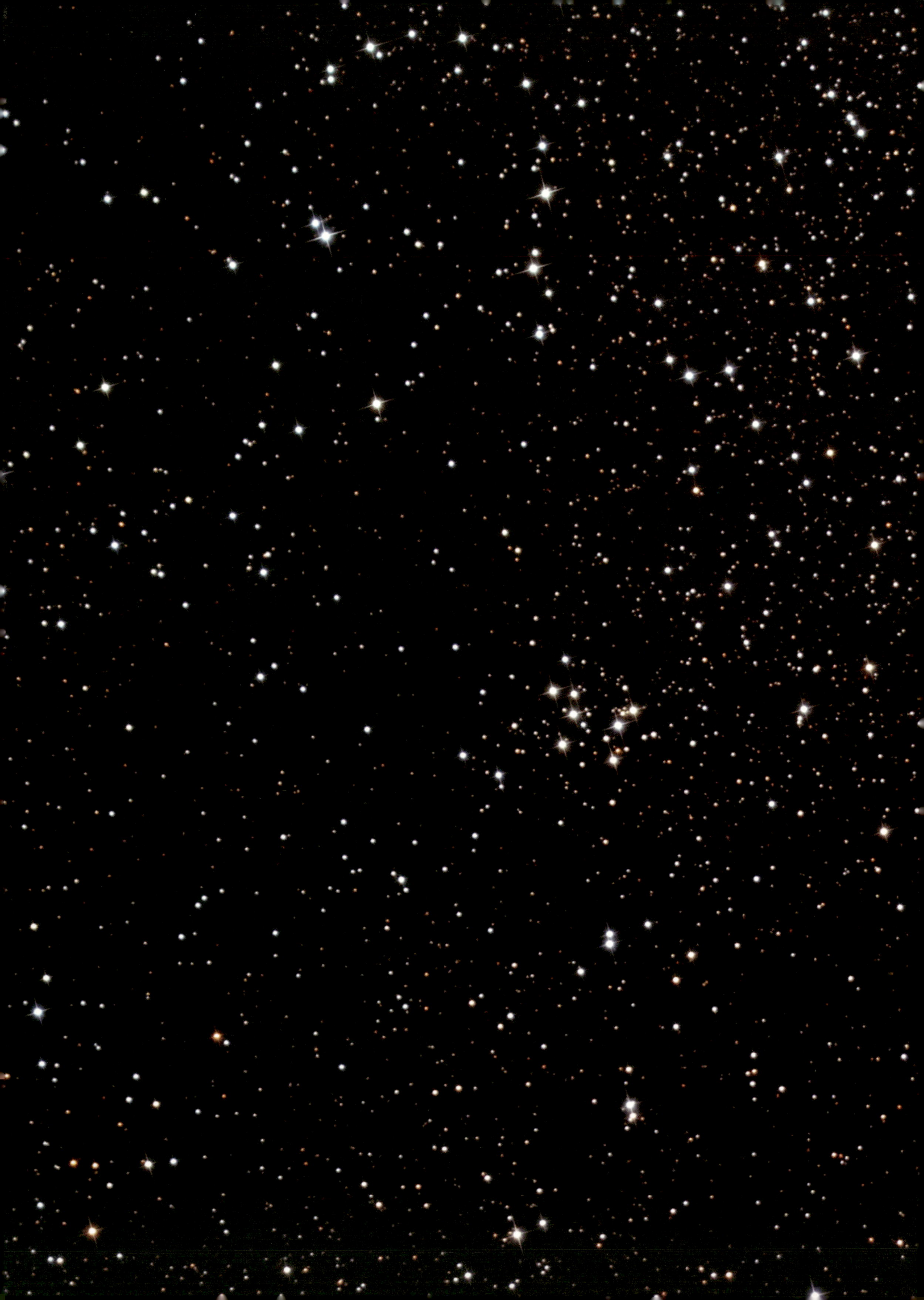

M29 – Open Cluster

This small open cluster can be found about 2 degrees south of Sadr, the central cross star in Cygnus. M29 lies in the Milky Way, so we see a lot of background stars in this image. Indeed, to the upper-right we see a greater density of stars, but the density diminishes as we look to the left. The reason for this is that there is a lot of intervening Milky Way dust that attenuates the light from the stars in this region; in fact astronomers estimate the light from the M29 cluster itself is dimmed by galactic dust by some three magnitudes! Although this little open cluster may not appear particularly spectacular in this image, it would be all the more so if it were three magnitudes brighter.

M29 image: 39 minutes exposure, SXV-H9C / Sky-90.

Quasar APM08279+5255

The light from this brilliant quasar, once misclassified as a red carbon star and since then heralded as the brightest known continuously emitting object (with an estimated energy output of 100 billion of our Suns), was recently studied by the Hubble Space Telescope. The Hubble revealed that its light is lensed by an intervening gravitational field (possibly from a galaxy), resulting in its brightness being magnified. Nonetheless, its extreme red shift, 3.87, places APM08279+5255 very far across the universe indeed, at an estimated distance of almost 13 billion light-years. The light we now see was emitted close to the time of the dawn of the Universe, less than 1 billion years after the Big Bang.

Seen in our image as just a dim, reddish dot, it is the single most distant object that we have ever photographed.

APM08279+5255 image from 3 hours 26 minutes exposure time, SXV-H9C / Hyperstar.

NGC6888 – The Crescent Nebula

Red supergiant Wolf-Rayet star WR136 at the center of this nebula is blowing out a high energy stellar wind, which is now impacting slower moving material ejected from the star at an earlier period in its life when it was a red giant. The result is the spectacular shell of glowing gas that we see here. Scientists predict that WR136 will undergo a supernova explosion some time in the next million years.

This object is called the Crescent Nebula because when it is not exposed deeply or is seen under visual observation just the bright crescent-shaped outer shell can be seen.

Some people say deeply exposed images of this nebula remind them of the head of a hamster.

Crescent Nebula image: 7 hours 20 minutes RGB plus 2 hours Ha exposure, SXV-H9C / Hyperstar.

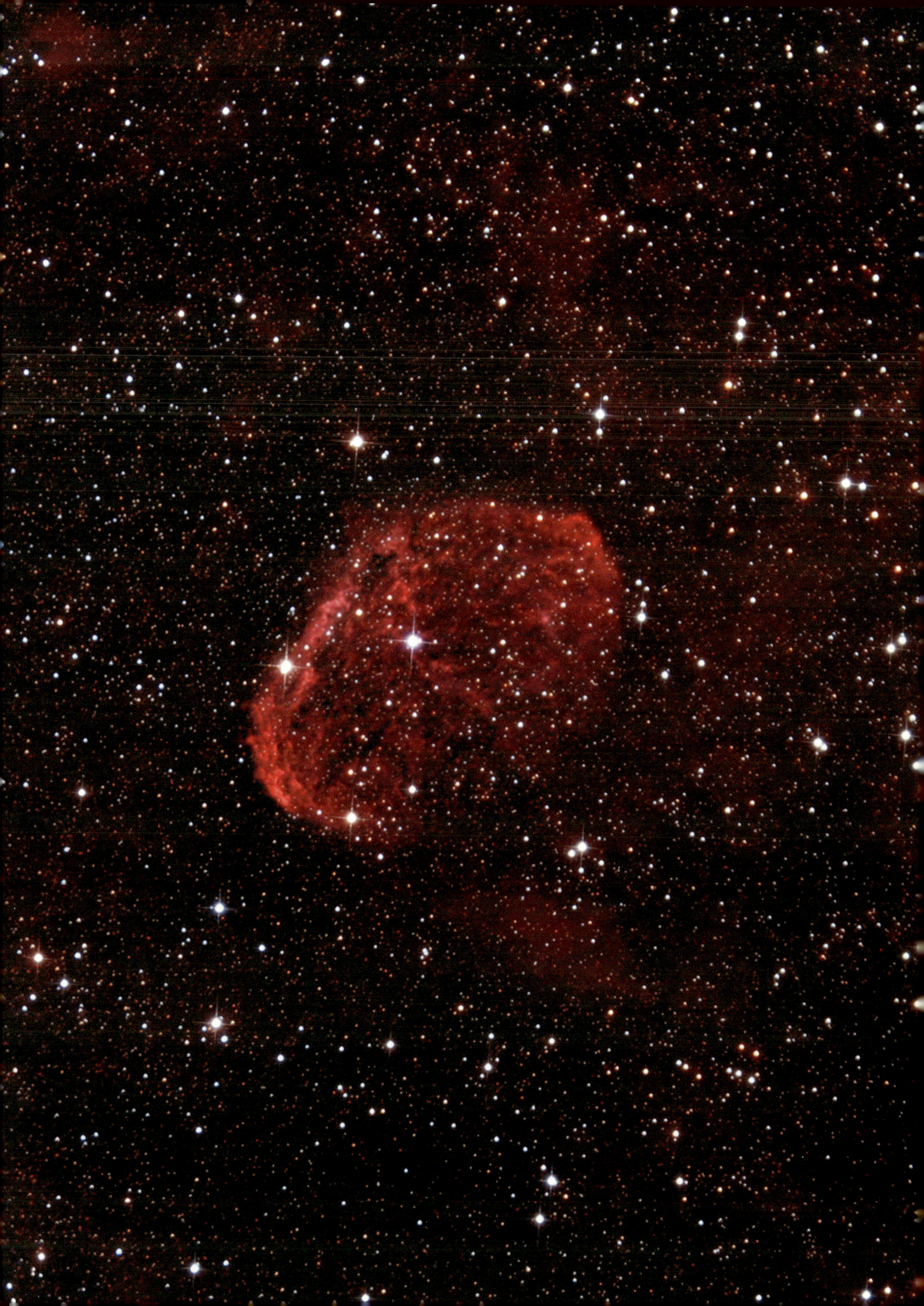

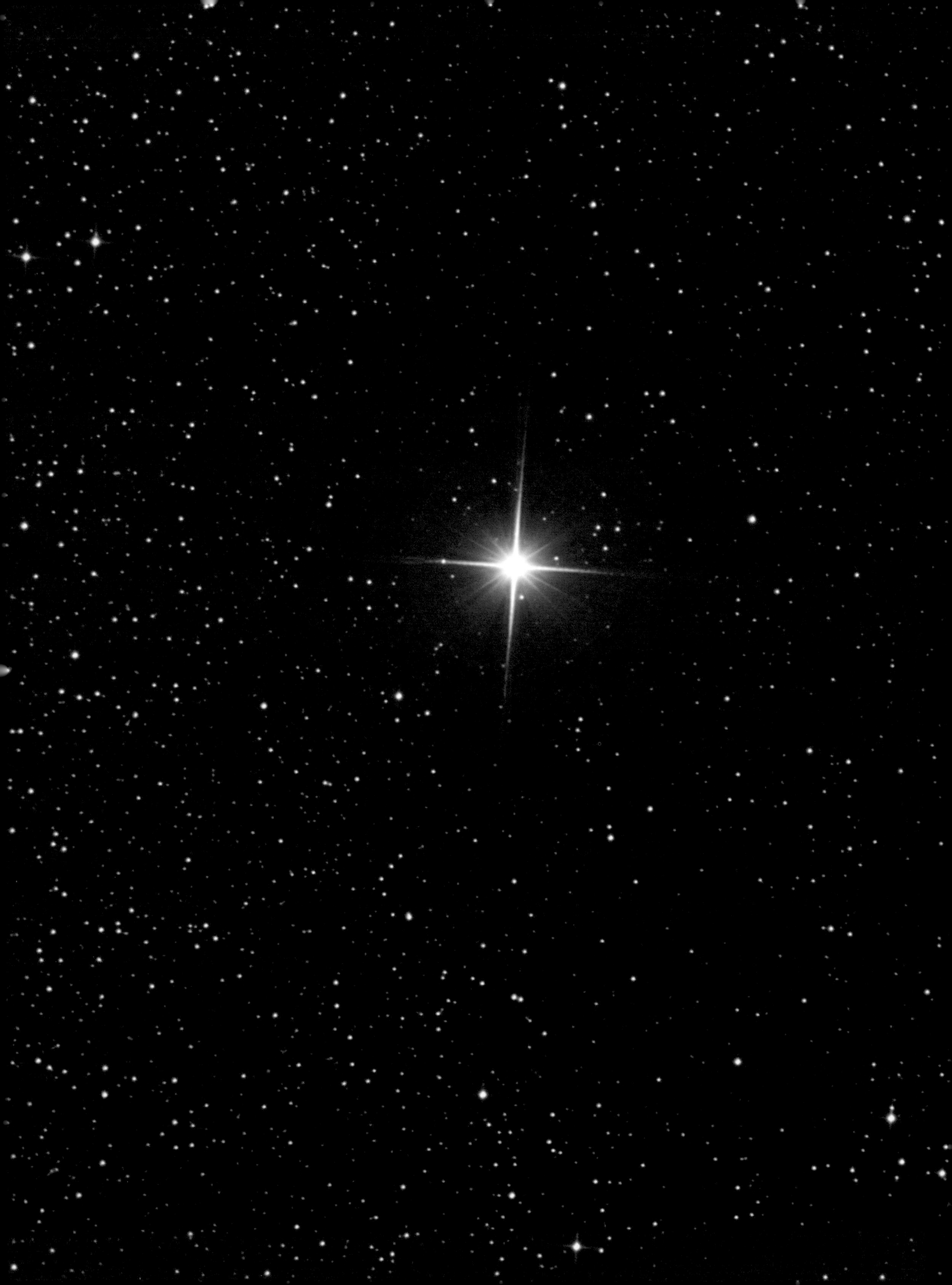

Deneb – Bright Star in Cygnus

The head of the Cygnus cross, and the 19th brightest star in our night sky, Deneb is one of the most luminous stars known in terms of absolute brightness. It is enormously brilliant - and simply enormous as well! Estimated at over 100,000 times the luminosity of our Sun, and 200 to 300 times the diameter, its surface would reach the orbit of Earth if it were in the position of our Sun. Deneb's mass is estimated at 20 to 25 times that of our Sun.

Our Sun's surface temperature is 5,800 degrees Kelvin. Blue-white Deneb, with its surface temperature a much hotter 9,000 degrees Kelvin, is emitting 5 times as much energy from each square meter of its much larger surface, consuming its nuclear fuel at such a rapid rate that it will last only a few million years before going supernova.

Deneb image: 24 minutes exposure time, SXV-H9C / Hyperstar.

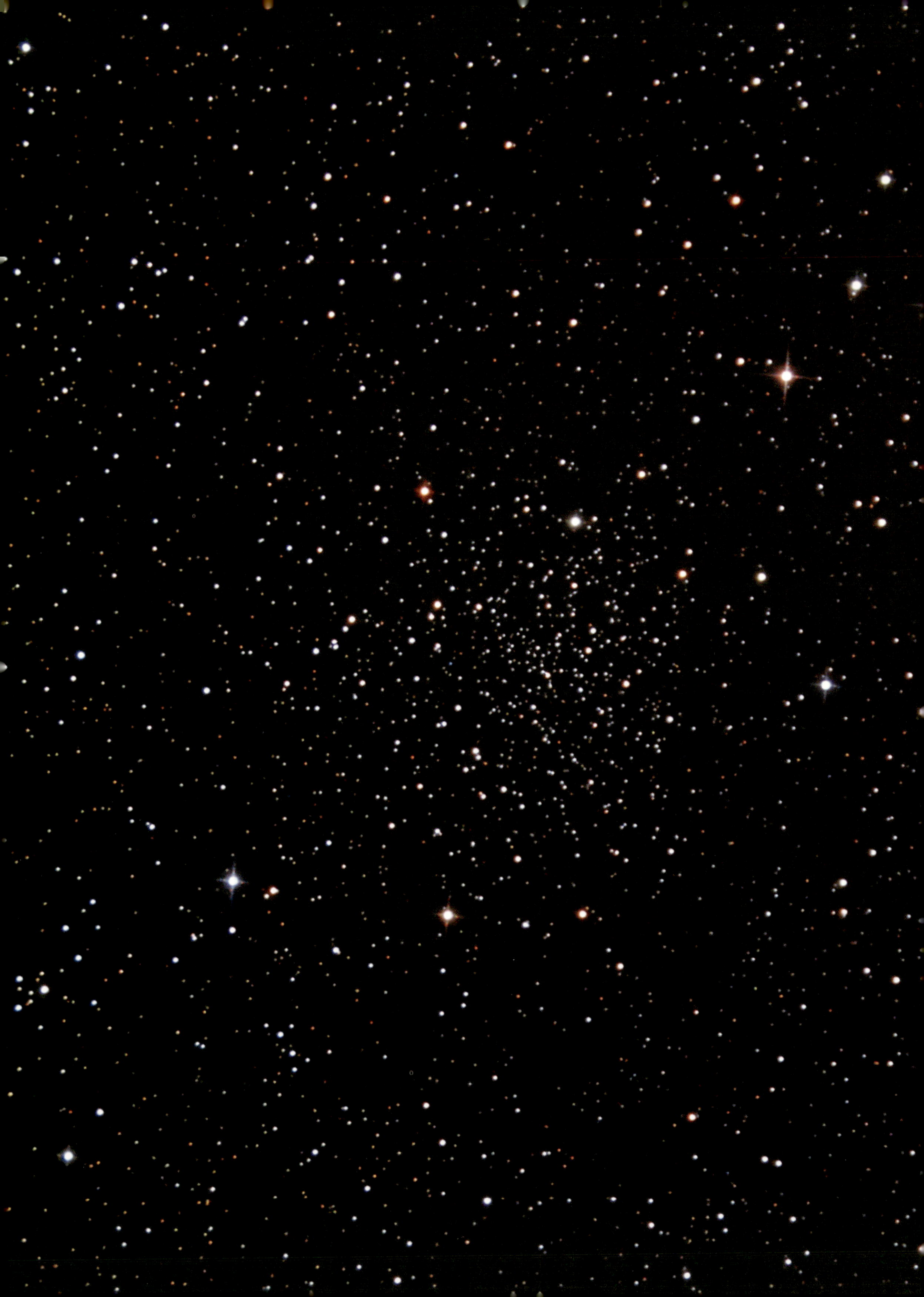

NGC188 – Open Cluster

First discovered in 1831 by John Herschel, this ancient and distant open cluster in Cepheus was ultimately cataloged as NGC188 in J. L. E. Dreyer's New General Catalog of 1888.

Not far from the celestial north pole, NGC188 is also designated Caldwell 1, the very first object in Sir Patrick Moore's catalog of 109 interesting deep-sky objects (known as the Caldwell objects).

Containing about 120 stars, some early estimates placed the age of the stars in this cluster at 24 billion years, which greatly exceeds what is now believed to be the entire age of the Universe! More recent estimates put it at a more conservative 12 to 14 billion years.

NGC188 image: 2 hours 20 minutes exposure time, SXV-H9C / Hyperstar

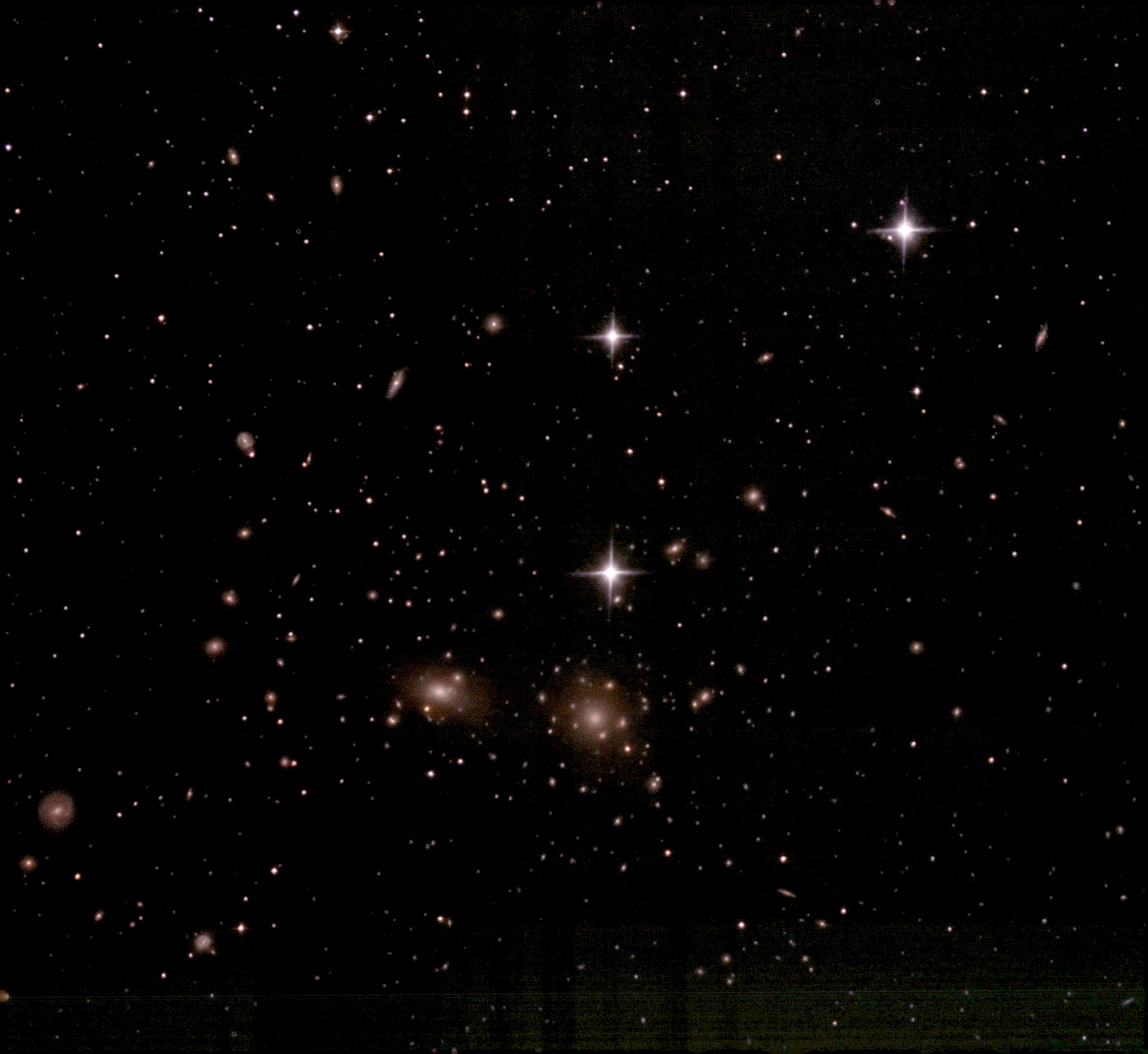

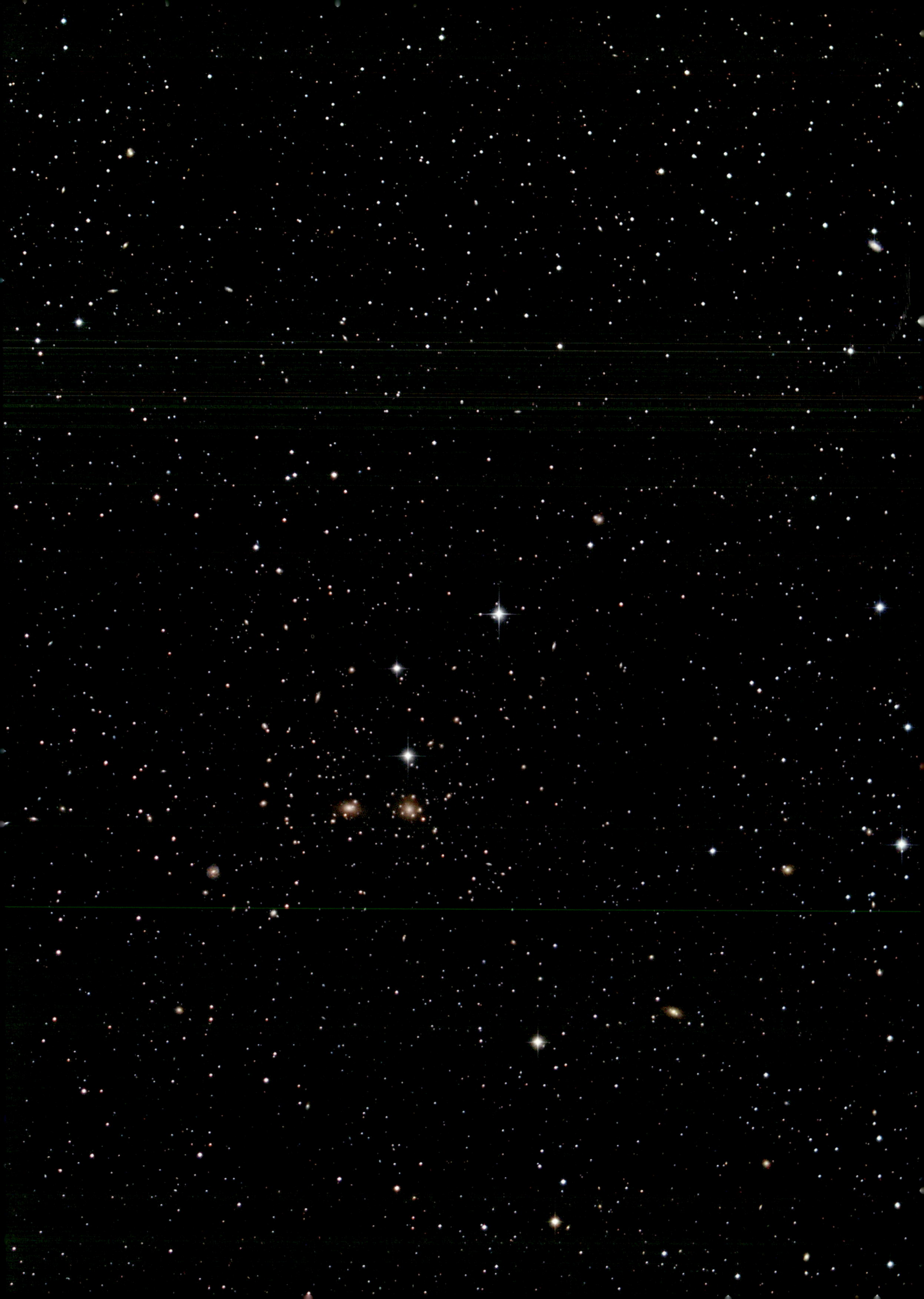

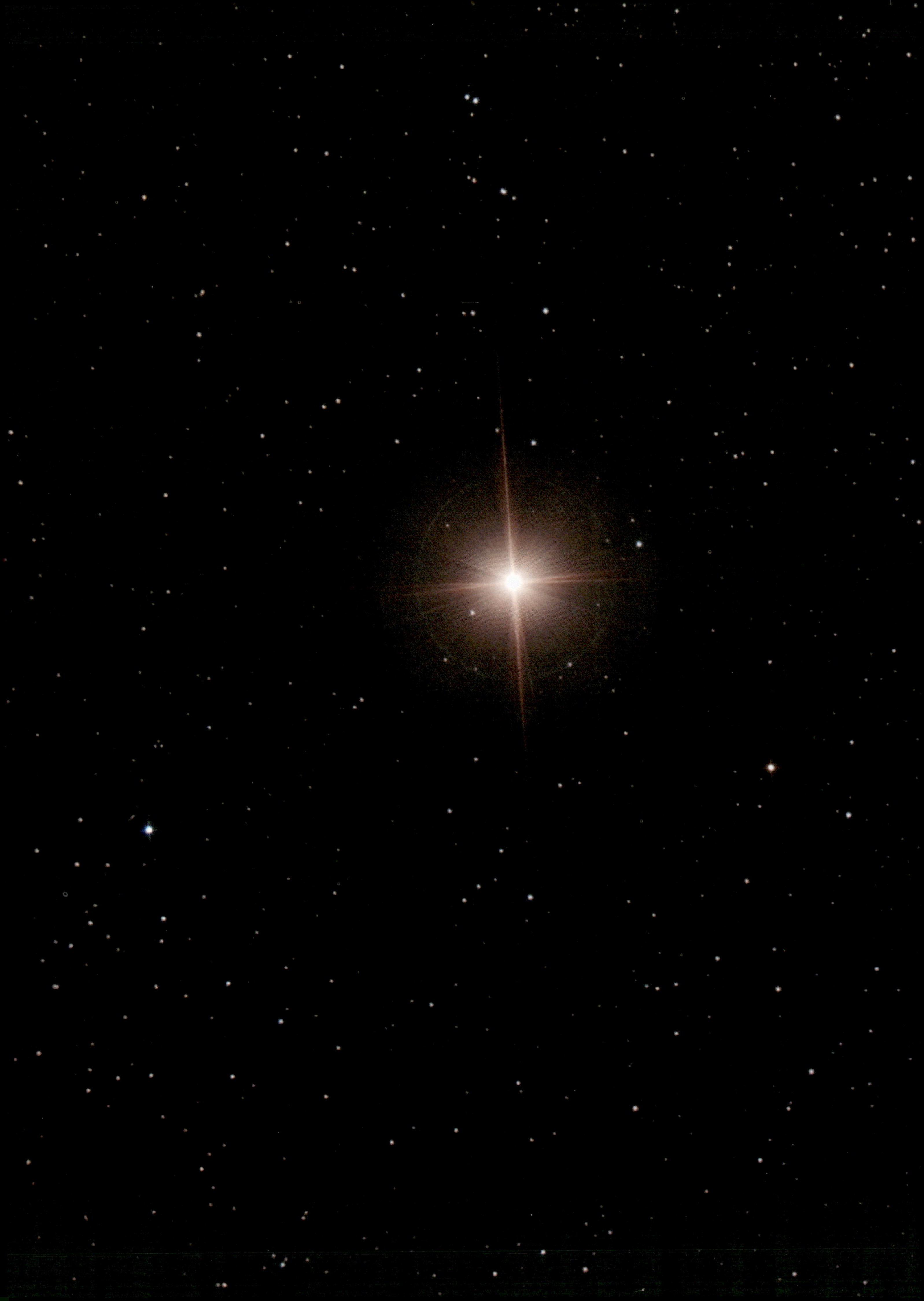

pposite

ldebaran – Bright Star in aurus

eye of the bull, Aldebaran is a very bright red giant star only 68 light-rs away in the constellation Taurus the Bull. The name Aldebaran is bic for "the follower," which refers to the way Aldebaran "follows" the ades open star cluster across the sky. It is the brightest star in Taurus, in fact one of the brightest stars in the entire night sky. Aldebaran's neter is 38 times that of the Sun, and it shines with 150 times the Sun's inosity.

1997 a team of astronomers announced the discovery of a substellar npanion (basically a giant planet) orbiting Aldebaran with a mass of 11 ter masses. This discovery remains unconfirmed.

2 million years NASA's *Pioneer 10* spacecraft will pass near Aldebaran. il then we will just have to enjoy from Earth the beauty of this large red-nge star in the constellation Taurus.

ebaran image: 24 minute exposure, SXV-H9C / Hyperstar.

Next Two Pages >

Rich Star Field in Cassiopeia

A deep exposure of the Milky Way in Cassiopeia reveals a dazzling array of stars, clusters of stars, and even some nebulosity.

Open clusters shown in this image include NGC663, NGC659, NGC654, Czernik 4, Trumpler 1, and M103 (also known as NGC581). The red nebula is LBN630.

Note also the apparent appearance of dark Milky Way dust clouds crossing the entire image – seen as a slight lessening of the star density.

Star field image: 6 hours 8 minutes RGB exposure, SXVF-M25C / Sky-90.

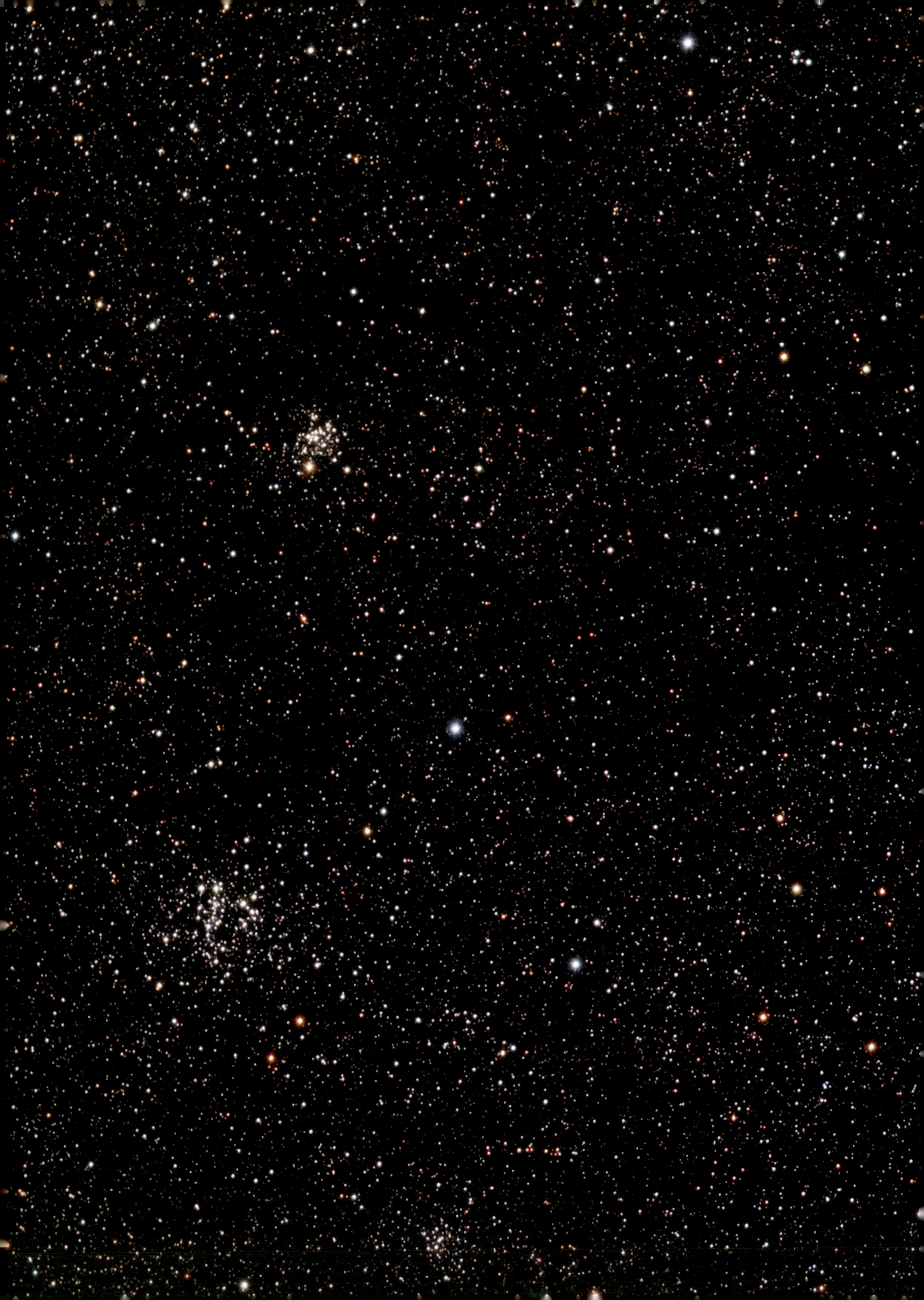

M105 and Other Galaxies in Leo

The fuzzy objects in this image are galaxies in central Leo. Top to bottom, the seven largest/brightest galaxies we can see in this image are NGC3377, NGC3367, NGC3412, NGC3384, M105 (also known as NGC3379), NGC3389 (the bluish one), and M96 (aka NGC3368).

The Leo I or M96 group of galaxies lie about 38 million light-years distant. NGC3389 is generally not considered a member of this group but rather a background object, as it is receding from us at a much greater speed than the rest.

M105 image: 3 hours RGB exposure, SXVF-M25C / Sky-90.

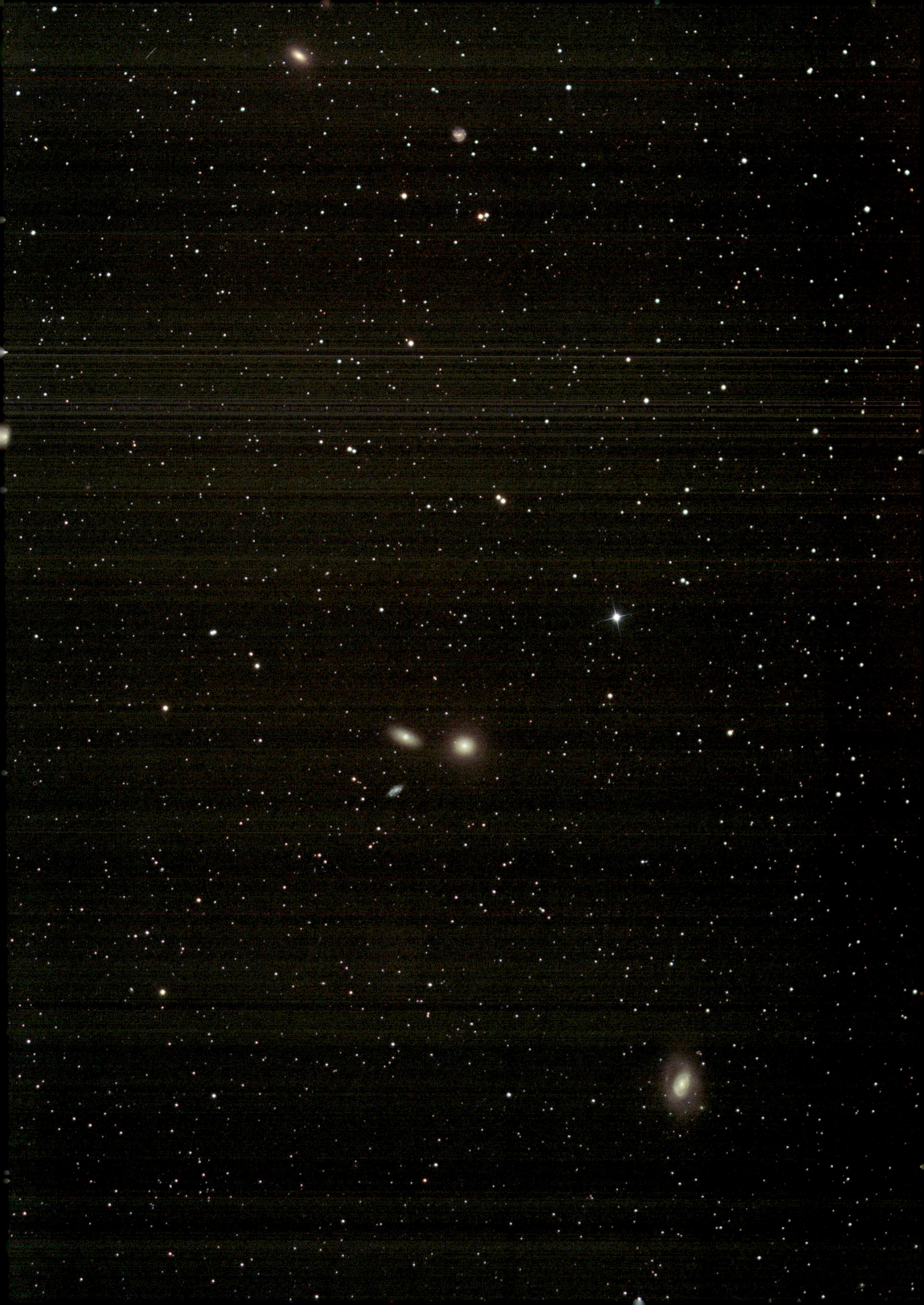

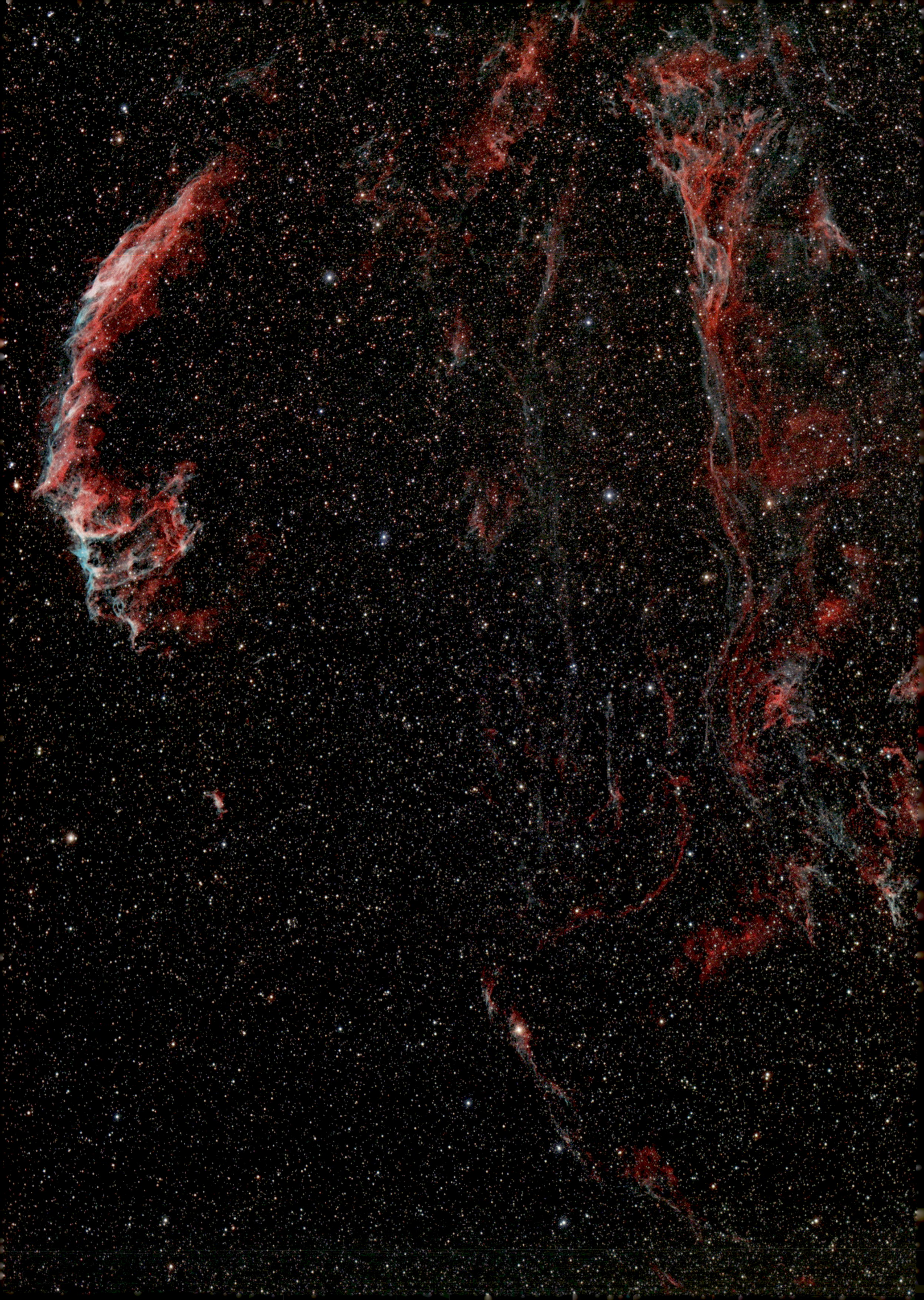

The Veil Nebula Complex

The Veil Nebula, the Network Nebula, the Cygnus Loop, the Bridal Veil Nebula, Caldwell 33 and 34, NGC6992/6995/6960, and IC1340 – all make up this complex of beautiful, colorful, filamentary Milky Way nebulae in the constellation Cygnus.

This entire complex is a supernova remnant that lies approximately 2,500 light-years from Earth. The supernova explosion that created this beautiful cosmic work of art occurred some 10,000 years ago and may have been witnessed by our primitive ancestors, to whom it must have looked as bright as the Moon. This 2-frame mosaic measures approximately 3.5 by 3 degrees, where the diameter of a full Moon is only half a degree across; in other words, the Veil Nebula complex is huge, though too dim to be seen by the unaided eye.

Veil Nebula Complex image assembled as a mosaic of several overlapping SXVF-M25C / Sky-90 images, total exposure time 32 hours in RGB, Ha, and OIII, and a similar amount of digital processing time.

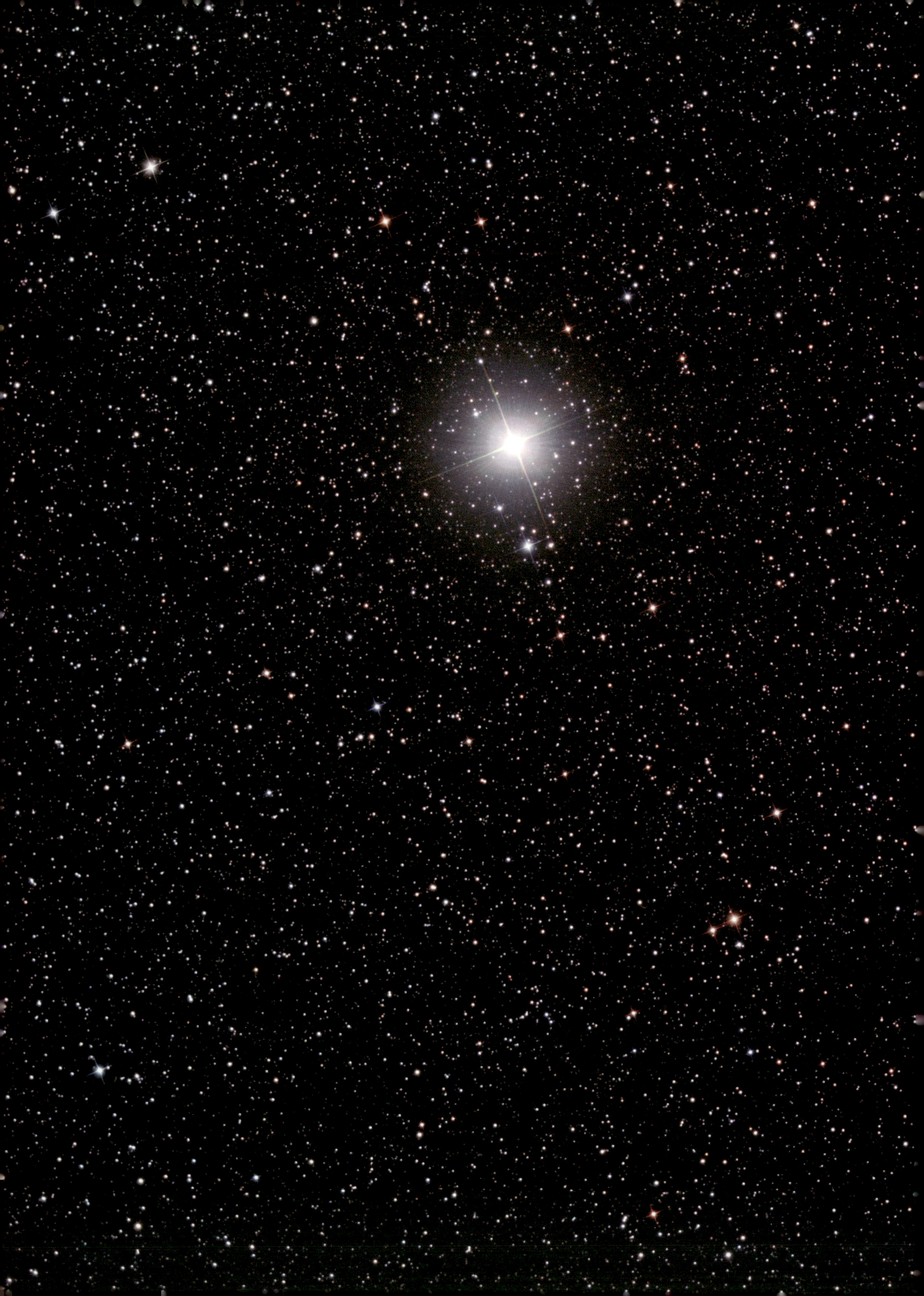

Procyon – Bright Star in Canis Minor

Shown here against a rich star field, Procyon is the 7th brightest star in the nighttime sky. It precedes the march of the brightest star, Sirius (not shown here), across the heavens from east to west, due to Earth's rotation.

Shining with an apparent magnitude of 0.4, this white F5-type star is only a little larger than our own Sun, but it appears very bright to us because it is relatively close by, at only 11.41 light-years distant. Procyon has a dim white dwarf companion star, Procyon B (not shown here, as it is very close to Procyon and is lost in the glare), which follows an oblong orbit around Procyon. As with many companions, the existence of Procyon B was inferred from carefully measured astrometric data before it was directly observed for the first time in 1896.

Procyon forms one of the three points of the Winter Triangle asterism, with Sirius and Betelgeuse as the other two points.

This is the first "single bright star" image we made with the wide-field SXVF-M25C / Sky90 camera/telescope combination. Total exposure time 3 hours 45 minutes.

Cone Nebula, Christmas Tree Cluster, Fox Fur Region

There is a great deal to see in this deep view into the Milky Way in the constellation Monoceros. Red glowing hydrogen contrasts with a blue-white reflection nebula, dark nebulae, and multicolored stars.

The central bright cluster of stars, NGC2264, is called the Christmas Tree Cluster. Above this cluster is the dark Cone Nebula, and slightly below that is the Fox Fur Nebula. There appears to be a comet towards the top-left of the image, but that is not what the object is; it is in fact Hubble's Variable Nebula, NGC2261. At the left edge of the image is a beautiful golden-colored open cluster, Trumpler 5.

This image from over 7 hours of RGB exposure time, SXVF-M25C / Sky-90.

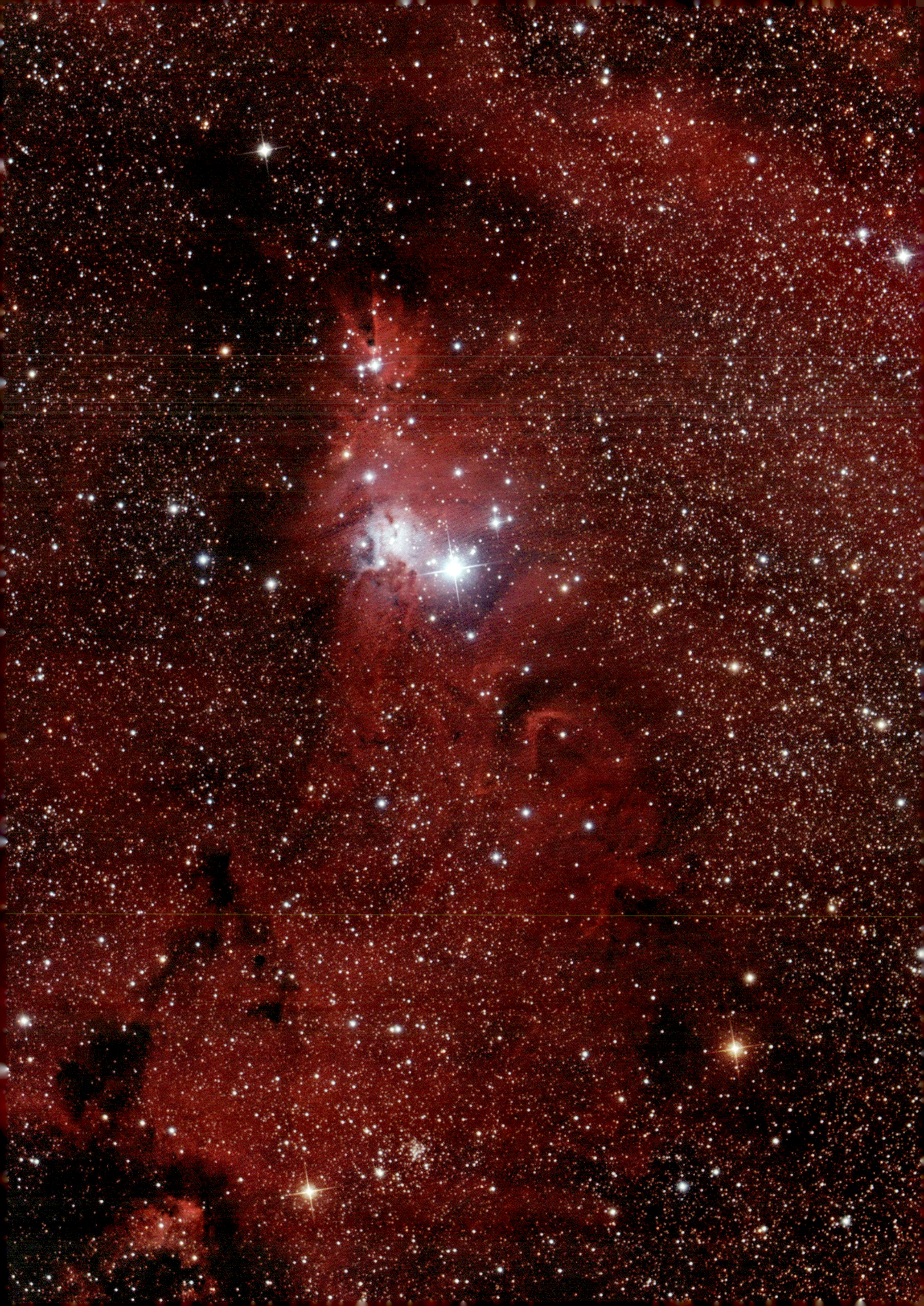

Shown in two views here at different magnifications, this beautiful spiral galaxy is clearly seen from almost directly above the plane and center of its rotation. Its spiral shape is so prominent that in 1851 it was one of the very first objects identified as a spiral nebula. That was long before astronomers realized there could be galaxies outside our own. M101 has also been called the Grand Design Spiral Galaxy. Several other smaller or more distant galaxies can also be seen in the wide-field view at right.

M101 image below: 1 hour 37 minutes RGB exposure time SXV-H9C/Hyperstar; image at right: 4 hours 39 minutes RGB exposure time, SXVF-M25C / Sky-90.

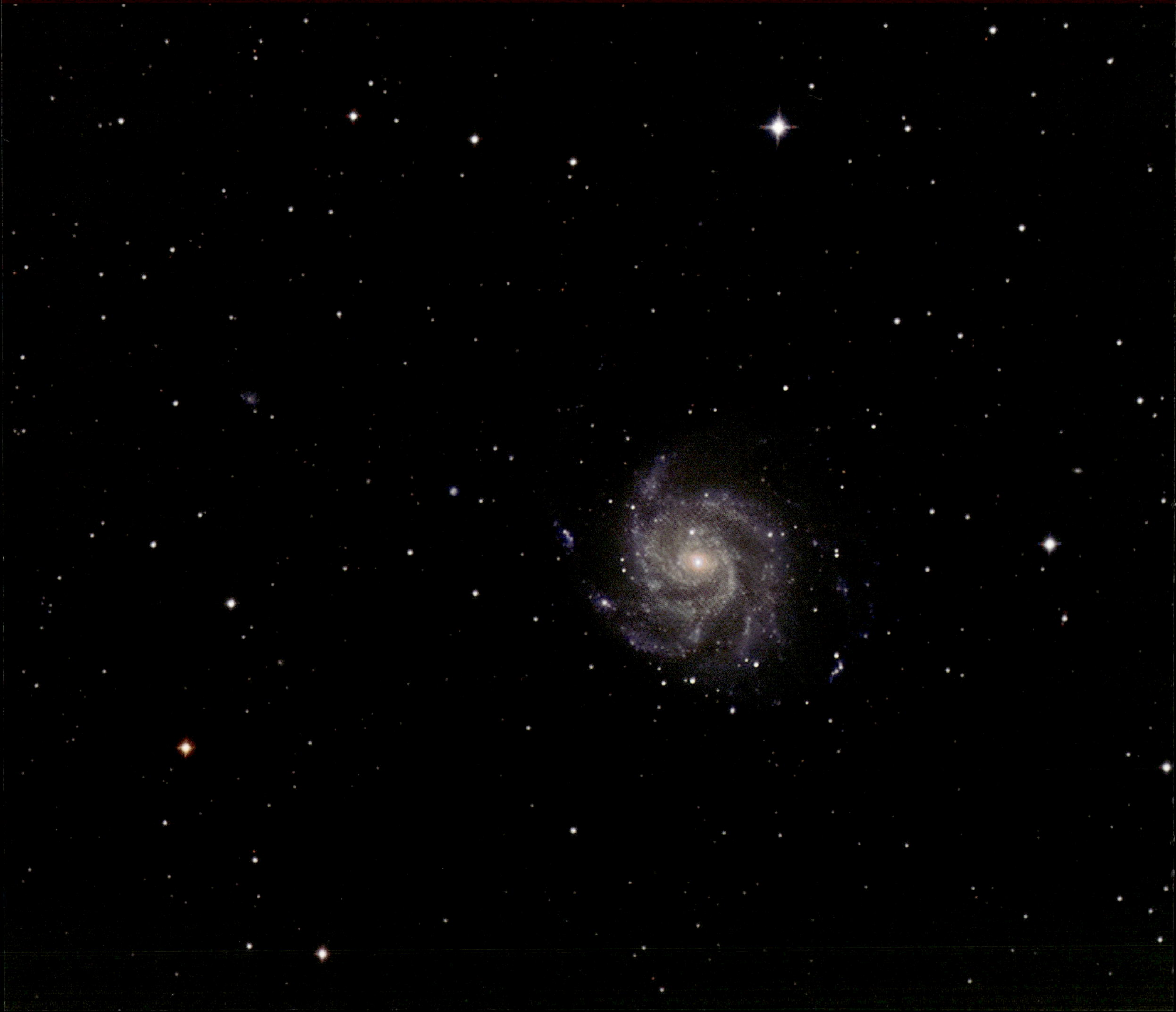

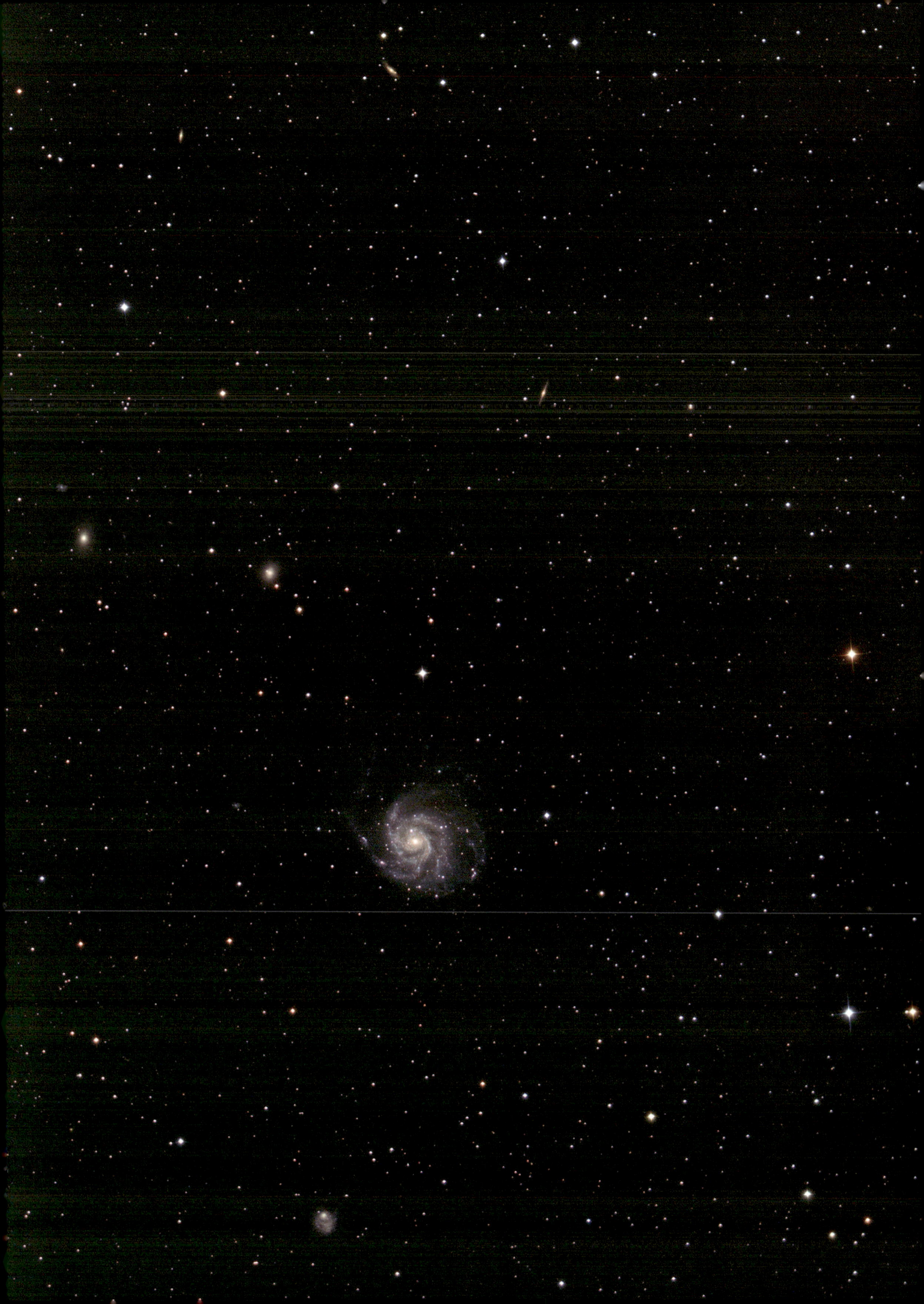

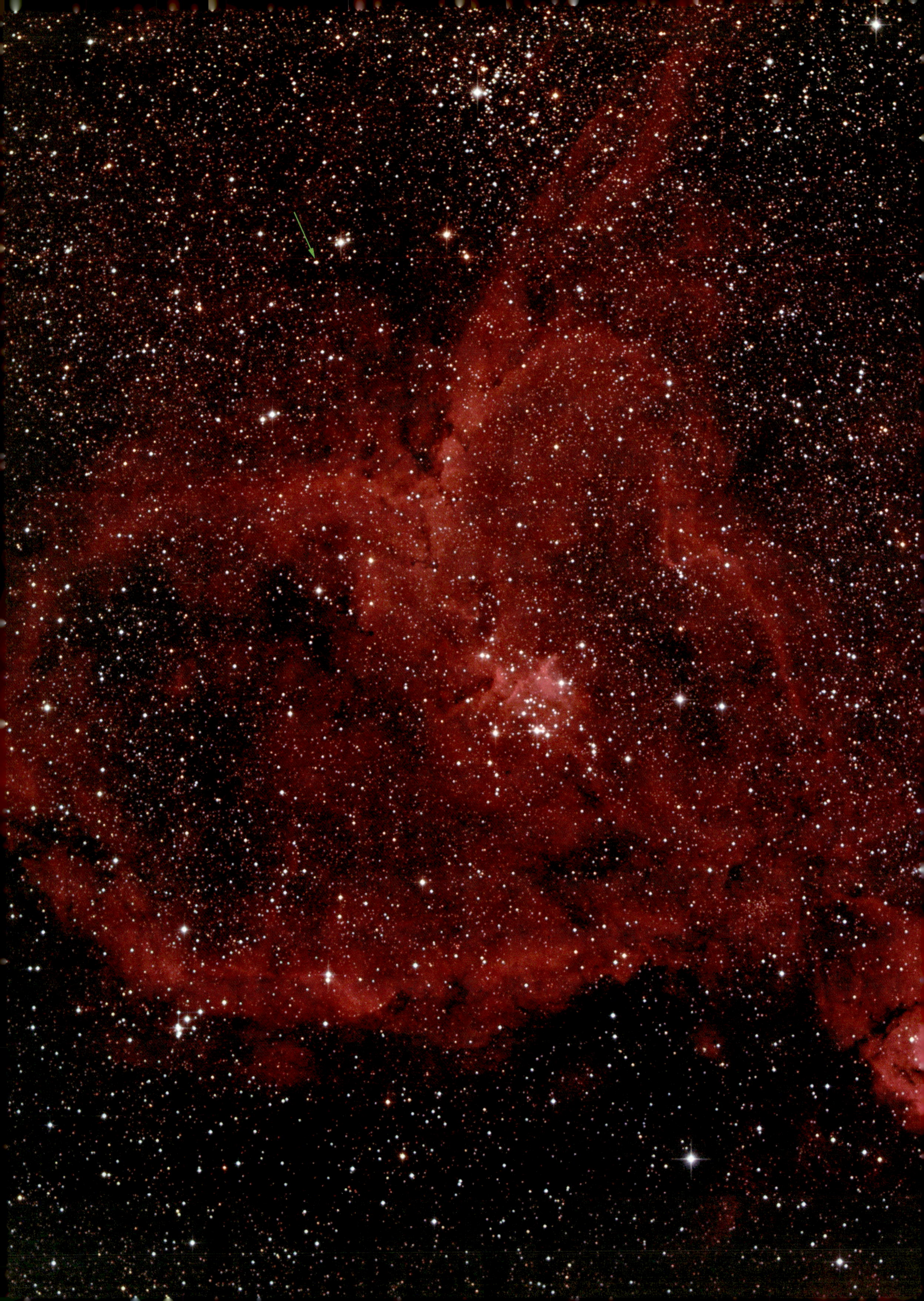

IC1805 – The Heart Nebula

Glowing hydrogen gas powered by a small central group of bright stars forms a cosmic shape reminiscent of a human heart in the constellation Cassiopeia. At 7,500 light-years distant from Earth, the Heart Nebula spans a distance of some 180 light-years.

An astronomical oddity has been found in this area – a microquasar designated LSI +61 303. This strange binary star emits very high-energy gamma rays. Astronomers believe it was expelled out the "top" of the heart millions of years ago as a result of a supernova explosion. Its position is marked on the left with a small arrow.

Below, the stars have been digitally removed; we can see the hydrogen emissions alone. This helps reveal the extent and complexity of the nebula.

Heart Nebula images from 6 hours RGB exposure and 1 hour 23 minutes Ha exposure, SXVF-M25C / Sky-90.

Capturing the Images

The images in *Star Vistas* fall broadly into two groups. The first group was acquired with a Starizona Hyperstar-equipped Celestron 11" C11 Schmidt-Cassegrain telescope and Starlight Xpress SXV-H9C 1.4 megapixel one-shot color camera with Hutech IDAS light pollution filter. The field of view of this optical system is one-degree by three-quarters of a degree, which is large enough for most "medium-sized" deep-sky objects. The main advantage of the Hyperstar/11 system was its incredible light-gathering capacity (photographically referred to as "speed") at f/1.85, which meant that exposures were typically only about a minute long. This means it was a simple matter to collect 100 sub-exposures in a single imaging session, which after stacking (a term describing the combining of multiple images of the same object) gave a very high signal-to-noise ratio in the final processed image. In layman's terms, images processed from 100 or more sub-exposures look clean and smooth!

The second group of images was captured with a different camera and telescope combination, a Starlight Xpress SXVF-M25C 6 megapixel one-shot color camera mounted to a Takahashi Sky-90 refractor telescope with an f/4.5 focal length reducer and Hutech IDAS light pollution filter, providing a much wider field of view at 3.33 by 2.22 degrees. This enormous field of view [FOV] allows one to capture the whole of the Andromeda Galaxy [M31] in a single frame. Of course some of the larger astronomical objects are still too big even for this setup! Some examples where multiple frames (mosaics) with the Sky-90/M25C are needed include the North America and Pelican nebulae region in Cygnus, the massive Gamma Cygni nebulosity surrounding the bright star Sadr (also in Cygnus), the Veil Nebula in Cygnus, and the massive IC1396 nebulosity in Cepheus.

Several images were taken using the SXV-H9C one-shot color camera on the Takahashi Sky 90 refractor; this was the interim imaging system used before acquiring the newer SXVF-M25C one-shot color camera. Images in this category include the double cluster in Perseus, NGC6914 in Cygnus, and the little open cluster, M29, in Cygnus.

In a nutshell, the Sky-90 telescope is mounted piggy-back on the C11, and an autoguider system is employed under computer control to ensure the telescopes track the movement of objects across the sky perfectly. Greg uses a software package called Maxim DL, which not only manages the guiding but also triggers the exposures in the camera and gathers the data. Once many exposures, each time-limited by the light pollution in the sky over the New Forest Observatory, have been captured, Greg derives the color information, stacks them in Maxim DL to average out noise and inaccuracy, then sends the resultant file to Noel in standard FITS format.

Processing the Images

Astro image processing can be intensive work, demanding much of the computer as well as the individual running it. Since the most important factor in acquiring and processing astro images is that we are doing everything we can to overcome limitations (in seeing through the atmosphere, in the optics, in the noise resistance and dynamic range in the cameras, data loss during processing, etc.), the computer systems have to be reliable, fast, and capable of handling very large amounts of information.

Noel uses a 64-bit Dell Precision workstation with dual processors, a fast Raid disk array, 8 gigabytes of RAM, and two large, sharp, top-quality LCD monitors. This workstation is quick, capable of working on huge images, and rock solid reliable.

Once Noel receives and saves the stacked files containing combined but otherwise unprocessed astro data from Greg, he starts by opening each file into Photoshop through a plug-in called FITS Liberator, which is available online at the NASA/ESO website. This plug-in provides Noel with both the ability to read standard astronomical FITS format data and to apply mathematical "developing" functions, which set initial brightness and contrast levels for the image. Noel uses one of his own Astronomy Tools actions to drive the plug-in to read in and assemble the three channels of information (red, green, and blue data).

Once the initial image is loaded into Photoshop, a typical image processing session involves digital removal of the level of light pollution-induced "sky fog" in the images; flattening the luminance levels (i.e., removing vignetting); further adjustment of "curves," which define the brightness and contrast of the image; defect removal (i.e., detection and removal of hot pixels, satellite trails, etc.); and final color balancing and detail enhancement for publication. All of the work is done at or above the camera's native resolution, so as not to lose important detail during processing. Many of the steps are automated through the use of Noel's Astronomy Tools actions, many of which he developed specifically to aid in the preparation of images for this book.

Typically the processing of an astro image can take as long as the acquisition of the data – often several hours or more. Combining exposures from multiple imaging sessions also multiplies the work and involves careful alignment and care to ensure seams are not visible.

Once an image is essentially ready for presentation – be it for viewing on the computer or printing or inclusion in a book – Noel sends it back to Greg for a final review. Only when both agree the very best possible result has been obtained is the image used.

that electronics engineers use electrons in conventional electronic circuits. So basically, he loves light and everything to do with light. Optical instruments, photography (high-speed with flashguns, or low-speed), lasers, LEDs, photonic crystals, lenses, prisms, spectrometers, and diffraction gratings – all are of equal interest.

Greg's fascination with photography is probably genetic, passed down by his father, who for some unknown reason, decided he would carry out plate (!) photography during World War 1 in the deserts of Egypt and Afghanistan. By the way, he did this all from horseback, as he was in the Lancers. He brought back many fascinating pictures of the northwest frontier, and was even in Luxor when Carter discovered Tutankhamen's tomb. With this sort of excitement in the Parker household it is perhaps not too surprising that the oldest boy (Greg's brother) spent his working life as a forensic photographer with Scotland Yard, and the youngest boy (Greg) took to high-speed and deep-space photography – the genes were just too strong.

Greg has published a textbook on semiconductor device physics, a book on how to take deep-space images [Springer], and the large-format book you currently hold in your hands. In addition, he has published over 120 scientific papers, a number of book chapters, and has over a dozen patents to his name. Finally, he has created four successful companies.

However, when all is said and done, his one true love is astrophotography, and if the night is clear he will be out in his New Forest Observatory capturing more ancient photons from the night skies.

- Greg Parker, Ph.D., C.Phys. C.Eng. F.Inst. P.
Brockenhurst, U.K. 2008

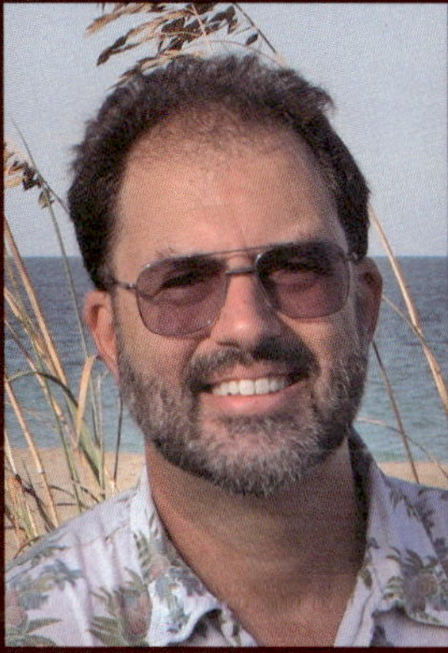

Noel Carboni is a software engineer and entrepreneur who spent his career developing reliable and high performance software and having recently founded ProDigital software, his own company. He loves photography, art, and astronomy.

Having taken on a role as a webmaster in the 1980s when the web was still young, he became involved with computer image processing and was introduced to Adobe Photoshop, then in its early versions. From that time forward he has developed his prowess with Photoshop and all things digital and graphic. When the software to do something he wanted to do was unavailable, he wrote a new program or action, often pioneering new methods. Over time, he developed a library of techniques for producing best-in-show digital imagery, and it dawned on him that the tools that enabled him to be so successful in processing images could be marketed to others. Now people worldwide use Noel's Photoshop add-on software to process their images. One of his most successful products is a set of actions called Astronomy Tools, which was used in preparing every image in this book.

Photography has been one of Noel's passions, and owning a great telescope was a life-long dream. He finally purchased a fine Meade instrument in 2004, coincidentally about the same time as Greg started in astrophotography. It followed that Noel connected his camera hardware to his new telescope, and he immediately produced ground-breaking lunar imagery. Unfortunately, the skies at Noel's home are bright with light pollution, and so he craved the ability to get great deep-sky image data that he could not capture himself. Thus was born the collaboration between Greg and Noel that led to this book.

Don't tell Noel something can't be done, because he'll be busy doing it.

- Noel Carboni
Florida, U.S., 2008

Index

Printed in the United States of America